Alijadallah Belabess

ADVANCED OLYMPIAD INEQUALITIES

Algebraic & Geometric Olympiad Inequalities

ISBN 978-1-7941-9392-5

Preface

I was fifteen years old back then, sitting in a café with my high school teacher. I still remember ordering orange juice while trying to solve very strange and challenging mathematical puzzles. My teacher was sitting to my left, reading the newspaper while sipping his black coffee. I always thought that I was really good at mathematics until that experience. My mind was buzzing with numbers, equations, and ideas, but I couldn't solve any of the problems. I drafted a couple of equations on the paper and told my teacher that I couldn't solve them. I still remember him saying to me, calmly but in a very clear and deep voice, "These are Olympiad problems... The only way to solve them is to think outside the box."

That was the beginning of my journey...

This story happened a long time ago, and since then, Olympiad problems have never failed to amaze me! There is a mysterious simplicity about them that makes them elegant and challenging at the same time. You might struggle with them for hours, but then you get a eureka moment when all the pieces come together in a pure moment of harmony.

In this book, I have collected some of my best inequalities which I have created over the years. I have tried to put them into a single location so that everyone can enjoy solving them as much as I enjoyed creating them. I tried to keep the inequalities simple but nevertheless sharp and elegant. I really hope that you can experience and enjoy their "Olympiadness".

Alijadallah Belabess
Khemisset, Morocco
π-day, 2019

Introduction

This book is designed for all students who, just like me, are fascinated by the amazing field of mathematical inequalities. I have been creating such problems for years now, and my drive has always been to challenge other mathematicians. I always try to find the right balance between simplicity, beauty, and difficulty. Although most of the problems could be solved using elementary techniques, they require some sort of imagination and creativity on the part of the reader. I have created more than a thousand inequalities throughout the years, and this book contains a selection of the best of them. The book is structured as follows:

Chapter 1 contains the most important inequalities that anyone who is aiming to participate in Olympiad competitions should be familiar with. The book does not include the proofs of these famous theorems, but it is straightforward to get them from other sources. I therefore strongly recommend the reader to study them thoroughly before moving on to the following chapters. This chapter contains four sections:

- Classical algebraic inequalities: this section covers most of the important theorems, such as Cauchy-Schwarz inequality, Chebyshev inequality, Schur's inequality, Power-Mean inequality, Jensen's inequality, and Rearrangement inequality in addition to other equally important but less known theorems such as Turkveich's inequality, Suryani's inequality, and Vasc's inequality. These theorems are often used as the building blocks in solving more complicated inequalities.

- Advanced algebraic techniques: this section covers more advanced topics, such as the UVW method, SOS theorem, Vasc's HCF, Mixing variables, the Lagrange Multipliers method, and SIP theorem. Most of these theorems could be proved using elementary tools and therefore could potentially provide valuable shortcuts during a mathematical contest.

- Useful inequalities: this section is a collection of some interesting and useful theorems that can be used to tackle harder problems (e.g., Nesbitt, Iran 1996). They were either given in Olympiad competitions around the world or posted online in mathematical forums. However, I believe that these inequalities are an integral part of the apparatus for proving inequalities of higher rank. This section could be of great benefit to anyone who is looking to expand his/her knowledge in the theory of inequalities.

- Geometric inequalities: this section covers the most important inequalities in the geometry of the triangle, such as Euler's inequality, Blundon's inequality, Gerresten's inequality, and Ciamberlini's inequality. Although these inequalities were discovered a long time ago, they are just as important today as they were in the past. Each of them will bring new insight to the reader on the laws and properties that govern our physical world. Furthermore, the reader will discover throughout this book that there is a natural equivalence between algebraic and geometric inequalities which can be used to unlock some difficult problems.

Chapters 2, **3**, and **4** contain an extensive collection of my own inequalities. They cover symmetric and non-symmetric inequalities, 3 and 4-variable inequalities alongside geometric inequalities. Each of these chapters ends with a selection of unsolved problems which were added as additional challenges to the readers.

Chapter 5 contains the solutions to the proposed inequalities. The solutions in this book are not necessarily mine; they are neither the shortest nor the smartest, but they are the ones I believe can be easily transposed to other problems. The chapter also contains a variety of "Comments" where I tried to either prove some famous inequalities used throughout this book or explain the rationale behind a specific approach.

Finally, I am very thankful to the people who enjoyed solving my inequalities on different platforms and forums and encouraged me to publish them. I am incredibly grateful for your support and trust. Thank you.

Contents

Chapter 1

Algebraic and Geometric Inequalities

1.1 Algebraic inequalities

1.1.1 Classical Algebraic inequalities

Theorem 1 (AM-GM). *Let $a_1, ..., a_n$ be non-negative real numbers. Then:*

$$\frac{a_1 + ... + a_n}{n} \geq \sqrt[n]{a_1...a_n}$$

with equality if and only if $a_1 = a_2 = ... = a_n$.

Theorem 2 (Cauchy-Schwarz). *Let $a_1, ..., a_n, b_1, ..., b_n$ be real numbers. Then:*

$$(a_1^2 + ... + a_n^2)(b_1^2 + ... + b_n^2) \geq (a_1 b_1 + ... + a_n b_n)^2$$

Theorem 3 (Titu). *Let $a_1, a_2, ..., a_n, b_1, b_2, ..., b_n$ be positive real numbers. Then:*

$$\frac{a_1^2}{b_1} + \frac{a_2^2}{b_2} + ... + \frac{a_n^2}{b_n} \geq \frac{(a_1 + a_2 + ... + a_n)^2}{b_1 + b_2 + ... + b_n}$$

Theorem 4 (Hölder). *Let $a_1, ..., a_n, b_1, ..., b_n$ be positive real numbers. Suppose that $p > 1$ and $q > 1$ satisfy $\frac{1}{p} + \frac{1}{q} = 1$. Then, we have:*

$$\left(\sum_{i=1}^{n} a_i^p\right)^{\frac{1}{p}} \left(\sum_{i=1}^{n} b_i^q\right)^{\frac{1}{q}} \geq \sum_{i=1}^{n} a_i b_i$$

More generally, let $x_{ij}(i = 1, ..., m, j = 1, ..., n)$ be positive real numbers. Suppose that $w_1, w_2, ..., w_n$ are positive real numbers satisfying $w_1 + ... + w_n = 1$. Then, we have:

$$\prod_{j=1}^{n} \left(\sum_{i=1}^{m} x_{ij}\right)^{w_j} \geq \sum_{i=1}^{m} \left(\prod_{j=1}^{n} x_{ij}^{w_j}\right)$$

Theorem 5 (Minkowski). *Let $a_1, ..., a_n$, $b_1, ..., b_n$ be positive real numbers. Suppose that $p > 1$. Then, we have:*

$$\left(\sum_{i=1}^{n} a_i^p\right)^{\frac{1}{p}} + \left(\sum_{i=1}^{n} b_i^p\right)^{\frac{1}{p}} \geq \left(\sum_{i=1}^{n} (a_i + b_i)^p\right)^{\frac{1}{p}}$$

Theorem 6 (Generalised Minkowski). *Let $a_{ij} \geq 0$ for $i = 1, ..., n$ and $j = 1, ..., m$ and let $p > 1$. Then:*

$$\left[\sum_{i=1}^{n}\left(\sum_{j=1}^{m} a_{ij}\right)^p\right]^{\frac{1}{p}} \leq \sum_{j=1}^{m}\left(\sum_{i=1}^{n} a_{ij}^p\right)^{\frac{1}{p}}$$

Theorem 7 (Chybeshev). *Let $a_1 \geq ... \geq a_n$ and $b_1 \geq ... \geq b_n$ be real numbers. Then:*

$$\frac{a_1 b_1 + ... + a_n b_n}{n} \geq \frac{(a_1 + ... + a_n)}{n}\frac{(b_1 + ... + b_n)}{n}$$

Theorem 8 (Rearrangement). *Let $a_1 \geq ... \geq a_n$ and $b_1 \geq ... \geq b_n$ be real numbers. For any permutation σ of $\{1, ..., n\}$, we have:*

$$\sum_{i=1}^{n} a_i b_i \geq \sum_{i=1}^{n} a_i b_{\sigma(i)} \geq \sum_{i=1}^{n} a_i b_{n+1-i}$$

Definition 1 (Convex Function). *Suppose that f is a one-variable function defined on $[a, b] \subset \mathbb{R}$. f is called a convex function on $[a, b]$ if and only if for all $x, y \in [a, b]$ and for all $0 \leq t \leq 1$, we have:*

$$tf(x) + (1 - t)f(y) \geq f(tx + (1 - t)y)$$

Theorem 9 (Jensen). *Let $f : [a, b] \to \mathbb{R}$ be a convex function. Then for any $x_1, ..., x_n \in [a, b]$ and non-negative real numbers $w_1, ..., w_n$ with $w_1 + ... + w_n = 1$, we have:*

$$\sum_{i=1}^{n} w_i f(x_i) \geq f\left(\sum_{i=1}^{n} w_i x_i\right)$$

Theorem 10 (Popoviciu). *Let $f : I \to \mathbb{R}$. If f is convex, then for any three points x, y, z in I:*

$$\frac{f(x) + f(y) + f(z)}{3} + f\left(\frac{x + y + z}{3}\right) \geq \frac{2}{3}\left[f\left(\frac{x + y}{2}\right) + f\left(\frac{y + z}{2}\right) + f\left(\frac{z + x}{2}\right)\right]$$

Definition 2 (Majorization). *Given two sequences $(a) = (a_1, a_2, ..., a_n)$ and $(b) = (b_1, b_2, ..., b_n)$ (where $a_i, b_i \in \mathbb{R}$ $\forall i \in \{1, 2, ..., n\}$). We say that the sequence (a) majorizes the sequence (b), and write $(a) \succ (b)$, if the following conditions are fulfilled:*

$$a_1 \geq a_2 \geq ... \geq a_n;$$
$$b_1 \geq b_2 \geq ... \geq b_n;$$
$$a_1 + a_2 + ... + a_n = b_1 + b_2 + ... + b_n;$$
$$a_1 + a_2 + ... + a_k \geq b_1 + b_2 + ... + b_k \quad \forall k \in \{1, 2, ..., n - 1\}.$$

Theorem 11 (Karamata). *Let $f : [a, b] \to \mathbb{R}$ be a convex function. Suppose that $(x_1, ..., x_n) \succ (y_1, ..., y_n)$ where $x_1, ..., x_n, y_1, ..., y_n \in [a, b]$. Then, we have:*

$$\sum_{i=1}^{n} f(x_i) \geq \sum_{i=1}^{n} f(y_i)$$

Theorem 12 (Weighted AM-GM). *Let $w_1, ..., w_n \geq 0$ such that $w_1 + ... + w_n = 1$. For all $x_1, ..., x_n > 0$, we have:*

$$\sum_{i=1}^{n} w_i x_i \geq \prod_{i=1}^{n} x_i^{w_i}$$

Theorem 13 (Schur). *Let x, y, z be non-negative real numbers. For any $r > 0$, we have:*

$$\sum_{cyc} x^r (x - y)(x - z) \geq 0$$

Theorem 14 (Generalised Schur). *Let a, b, c, x, y, z be six non-negative real numbers such that the sequences (a, b, c) and (x, y, z) are similarly sorted. Then:*

$$x(a - b)(a - c) + y(b - c)(b - a) + z(c - a)(c - b) \geq 0$$

Theorem 15 (Newton). *Let $x_1, ..., x_n$ be non-negative real numbers. Define the symmetric polynomials $s_0, s_1, ..., s_n$ by $(x + x_1)(x + x_2)...(x + x_n) = s_n x^n + ... + s_1 x + s_0$, and define the symmetric averages by $d_i = \frac{s_i}{\binom{n}{i}}$. Then:*

$$d_i^2 \geq d_{i+1} d_{i-1}$$

Theorem 16 (Maclaurin). *Let $x_1, ..., x_n$ be non-negative real numbers. Define the symmetric polynomials $s_0, s_1, ..., s_n$ by $(x + x_1)(x + x_2)...(x + x_n) = s_n x^n + ... + s_1 x + s_0$, and define the symmetric averages by $d_i = \frac{s_i}{\binom{n}{i}}$. Then:*

$$d_1 \geq \sqrt[2]{d_2} \geq \sqrt[3]{d_3} \geq ... \geq \sqrt[n]{d_n}$$

Theorem 17 (Muirhead). *Suppose that $(a_1, ..., a_n) \succ (b_1, ..., b_n)$, and $x_1, ..., x_n$ are positive real numbers, then:*

$$\sum_{sym} x_1^{a_1} x_2^{a_2} ... x_n^{a_n} \geq \sum_{sym} x_1^{b_1} x_2^{b_2} ... x_n^{b_n}$$

where the symmetric sum is taken over all $n!$ permutations of $(x_1, x_2, ..., x_n)$.

Theorem 18 (Power Mean). *Let $x_1, ..., x_n > 0$. The power mean of order r is defined by:*

$$\begin{cases} M_{(x_1, ..., x_n)}(0) = \sqrt[n]{x_1 ... x_n} \\ M_{(x_1, ..., x_n)}(r) = \left(\frac{x_1^r + ... + x_n^r}{n} \right)^{\frac{1}{r}} \quad r \neq 0 \end{cases}$$

Then, $M_{(x_1, ..., x_n)} : \mathbb{R} \to \mathbb{R}$ is continuous and monotone increasing.

Theorem 19 (Bernoulli). *For every real number $r \geq 1$ and real number $x \geq -1$, we have:*

$$(1+x)^r \geq 1 + rx$$

while for $0 \leq r \leq 1$ and real number $x \geq -1$, we have:

$$(1+x)^r \leq 1 + rx$$

Theorem 20 (Aczel). *Let $a_1, ..., a_n, b_1, ..., b_n$ be non-negative real numbers satisfying $a_1^2 \geq a_2^2 + ... + a_n^2$ and $b_1^2 \geq b_2^2 + ... + b_n^2$. Then:*

$$a_1 b_1 - (a_2 b_2 + ... + a_n b_n) \geq \sqrt{\left(a_1^2 - (a_2^2 + ... + a_n^2)\right)\left(b_1^2 - (b_2^2 + ... + b_n^2)\right)}$$

Theorem 21 (Suranyi). *For any non-negative real numbers $a_1, a_2, ..., a_n$, we have:*

$$(n-1)\sum_{i=1}^{n} a_i^n + n\prod_{i=1}^{n} a_i \geq \left(\sum_{i=1}^{n} a_i\right)\left(\sum_{i=1}^{n} a_i^{n-1}\right)$$

Theorem 22 (Shleifer). *For any non-negative real numbers $a_1, a_2, ..., a_n$, we have:*

$$(n-1)\sum_{i=1}^{n} a_i^4 + n\left(\prod_{i=1}^{n} a_i\right)^{\frac{4}{n}} \geq \left(\sum_{i=1}^{n} a_i^2\right)^2$$

Theorem 23 (Huygens). *Let $a_1, a_2, ..., a_n, b_1, b_2, ..., b_n, w_1, w_2, ..., w_n$ be positive real numbers such that $w_1 + w_2 + ... + w_n = 1$. Then:*

$$\prod_{i=1}^{n} (a_i + b_i)^{w_i} \geq \prod_{i=1}^{n} a_i^{w_i} + \prod_{i=1}^{n} b_i^{w_i}$$

Theorem 24 (Vasc). *Let a, b, c be non-negative real numbers. Then:*

$$(a^2 + b^2 + c^2)^2 \geq 3(a^3 b + b^3 c + c^3 a)$$

with equality at $a = b = c$ or $a : b : c = \sin^2 \frac{4\pi}{7} : \sin^2 \frac{2\pi}{7} : \sin^2 \frac{\pi}{7}$ and permutations.

Theorem 25 (Turkevich). *Let a, b, c, d be non-negative real numbers. Then:*

$$a^4 + b^4 + c^4 + d^4 + 2abcd \geq a^2 b^2 + b^2 c^2 + c^2 d^2 + d^2 a^2 + a^2 c^2 + b^2 d^2$$

Theorem 26 (Heinz). *For $a, b > 0$ and $\alpha \in [0, 1]$, we have:*

$$\sqrt{ab} \leq \frac{a^\alpha b^{1-\alpha} + a^{1-\alpha} b^\alpha}{2} \leq \frac{a+b}{2}$$

1.1.2 Advanced algebraic techniques

Theorem 27 (Maximum of Convex Functions). *Let $F(x_1, x_2, ..., x_n)$ be a real function defined on $[a, b] \times [a, b] \times ... \times [a, b] \subset \mathbb{R}^n$ with $a < b$ such that for all $k \in \{1, 2, ..., n\}$, if we fix $n - 1$ variables $x_j (j \neq k)$ then $F(x_1, x_2, ..., x_n) = f(x_k)$ is a convex function of x_k. F attains its maximum at the point $(\alpha_1, \alpha_2, ..., \alpha_n)$ if and only if $\alpha_i \in \{a, b\} \quad \forall i \in \{1, 2, ..., n\}$.*

Theorem 28 (uvw). *Assume that we are given the constraint $a, b, c \geq 0$. Let $a + b + c = 3u$, $ab + bc + ca = 3v^2$ and $abc = w^3$. Then the following is true:*

- *when we have fixed u, v^2 and there exists at least one value of w^3 such that there exist $a, b, c \geq 0$ corresponding to u, v^2, w^3: then w^3 has a global maximum and minimum. w^3 assumes a maximum value only when two of a, b, c are equal, and a minimum value either when two of a, b, c are equal or when one of them is zero.*

- *when we have fixed u, w^3 and there exists at least one value of v^2 such that there exist $a, b, c \geq 0$ corresponding to u, v^2, w^3: then v^2 has a global maximum and minimum. v^2 assumes a maximum value only when two of a, b, c are equal, and a minimum value only when two of a, b, c are equal.*

- *when we have fixed v^2, w^3 and there exists at least one value of u such that there exist $a, b, c \geq 0$ corresponding to u, v^2, w^3: then u has a global maximum and minimum. u assumes a maximum value only when two of a, b, c are equal, and a minimum value only when two of a, b, c are equal.*

Theorem 29 (SOS). *Consider the sum*

$$S = S_a(b - c)^2 + S_b(c - a)^2 + S_c(a - b)^2$$

then $S \geq 0$ if any of the following take place:

- $S_a, S_b, S_c \geq 0$ *are all non-negative.*

- $S_a, S_c, S_a + 2S_b, S_c + 2S_b$ *are all non-negative.*

- $S_b, S_b + S_a$ *and* $S_b + S_c$ *are all non-negative when b is the median of $\{a, b, c\}$.*

- S_b, S_c *and* $b^2 S_a + a^2 S_b$ *are all non-negative when $a \geq b \geq c$.*

- $S_a + S_b + S_c \geq 0$ *and* $S_a S_b + S_b S_c + S_c S_a \geq 0$.

Theorem 30 (Extreme Value). *If $f : X \to \mathbb{R}$ is real valued and continuous function from a compact space to the real numbers, then f attains a greatest value, that is there is an $x \in X$ such that $f(x) \geq f(y)$ for all $y \in X$.*

Theorem 31 (Lagrange Multipliers). *Let U be an open subset of $\mathbb{R}^n$, and let $f : U \to \mathbb{R}$ and $g : U \to \mathbb{R}$ be continuous functions with continuous first derivatives. Define the constraint set $S = \{x \in U | g(x) = c\}$ for some real number c. Suppose $x \in S$ is a local extrema of f, meaning x gives minimal or maximal value for some neighborhood around x. Then either $\nabla g(x) = 0$ at this point, or for some real number λ, $\nabla f(x) = \lambda \nabla g(x)$.*

Theorem 32 (Mixing Variables). *If $F : I \subset R^n \to R$ is a symmetric, continuous function and for all $(x_1, x_2, ..., x_n)$, we have:*

$$F(x_1, x_2, ..., x_n) \geq F\left(\frac{x_1 + x_2}{2}, \frac{x_1 + x_2}{2}, x_3, ..., x_n\right)$$

Then:

$$F(x_1, x_2, ..., x_n) \geq F(x, x, ..., x)$$

where $x = \frac{x_1 + x_2 + ... + x_n}{n}$.

Theorem 33 (Strong Mixing Variables). *If $F : I \subset R^n \to R$ is a symmetric, continuous function and for all $(x_1, x_2, ..., x_n)$, we have:*

$$F(x_1, x_2, ..., x_n) \geq F(y_1, y_2, ..., y_n)$$

where $(y_1, y_2, ..., y_n)$ is performed from $(x_1, x_2, ..., x_n)$ by Δ-transformation, then:

$$F(x_1, x_2, ..., x_n) \geq F(x, x, ..., x)$$

where $x = \frac{x_1 + x_2 + ... + x_n}{n}$.

Theorem 34 (Single Inflexion Point). *Let f be a twice differentiable function on R with a single inflexion point. For a fixed real number s, we denote:*

$$g(x) = (n-1)f(x) + f\left(\frac{s-x}{n-1}\right)$$

For all real numbers x_1, x_2, ..., x_n such that $x_1 + x_2 + ... + x_n = s$, we have:

$$\inf_{x \in R} g(x) \leq f(x_1) + f(x_2) + ... + f(x_n) \leq \sup_{x \in R} g(x)$$

Theorem 35 (Left Concave-Right Convex Function). *Let $a < c < b$ be real numbers, and let f be a continuous function on $I = [a, b)$, concave on $[a, c]$ and convex on $[c, b)$.*

If $x_1, x_2, ..., x_n \in I$ such that $x_1 + x_2 + ... + x_n = s$, where $s < (n-1)c + b$, then the expression:
$$E = f(x_1) + f(x_2) + ... + f(x_n)$$
is maximal for $x_1 = x_2 = ... = x_{n-1} \leq x_n$.

Theorem 36 (Right Convex Function). *Let f be a function on an interval $I \subset R$ and convex for $x \geq s$, $s \in I$. If*

$$f(x_1) + f(x_2) + ... + f(x_n) \geq nf\left(\frac{x_1 + x_2 + ... + x_n}{n}\right) \quad (*)$$

is true for all $x_1, x_2, ..., x_n \in I$ such that $x_1 + x_2 + ... + x_n = ns$ and $x_1 = x_2 = ... = x_{n-1} \geq s$, then $(*)$ is true for all $x_1, x_2, ..., x_n \in I$ such that $x_1 + x_2 + ... + x_n \geq ns$.

Theorem 37 (Half Convex Function). *Let $f(u)$ be a function defined on a real interval I and convex on $I_{u \geq s}$ or $I_{u \leq s}$ where $s \in I$. The inequality:*

$$f(a_1) + f(a_2) + ... + f(a_n) \geq f\left(\frac{a_1 + a_2 + ... + a_n}{n}\right)$$

holds for all $a_1, a_2, ..., a_n \in I$ satisfying $a_1 + a_2 + ... + a_n = ns$ if and only if

$$f(x) + (n-1)f(y) \geq nf(s)$$

for all $x, y \in I$ such that $x + (n-1)y = ns$.

Theorem 38 (Equal Variable Method). *Let $a_1, a_2, ..., a_n$ $(n \geq 3)$ be fixed non-negative real numbers, and let $0 \leq x_1 \leq x_2 \leq ... \leq x_n$ such that:*

$$x_1 + x_2 + ... + x_n = a_1 + a_2 + ... + a_n$$

$$x_1^p + x_2^p + ... + x_n^p = a_1^p + a_2^p + ... + a_n^p$$

and let $E = x_1^q + x_2^q + ... + x_n^q$.

- *Case 1: $p \leq 0$ (with $p = 0$ yields $x_1 x_2 ... x_n = a_1 a_2 ... a_n > 0$).*

(a) For $q \in [p, 0] \cup [1, +\infty[$, E is maximal when $0 < x_1 = x_2 = ... = x_{n-1} \leq x_n$, and is minimal when $0 < x_1 \leq x_2 = x_3 = ... = x_n$.

(b) For $q \in]-\infty, p] \cup [0, 1]$, E is minimal when $0 < x_1 = x_2 = ... = x_{n-1} \leq x_n$, and is maximal when $0 < x_1 \leq x_2 = x_3 = ... = x_n$.

- *Case 2: $0 < p < 1$.*

(a) For $q \in [0, p] \cup [1, +\infty[$, E is maximal when $0 \leq x_1 = x_2 = ... = x_{n-1} \leq x_n$, and is minimal when either $x_1 = 0$ or $0 < x_1 \leq x_2 = x_3 = ... = x_n$.

(b) For $q \in]-\infty, 0] \cup [p, 1]$, E is minimal when $0 \leq x_1 = x_2 = ... = x_{n-1} \leq x_n$, and is maximal when either $x_1 = 0$ or $0 < x_1 \leq x_2 = x_3 = ... = x_n$.

- *Case 3: $p > 1$.*

(a) For $q \in [0, 1] \cup [p, +\infty[$, E is maximal when $0 \leq x_1 = x_2 = ... = x_{n-1} \leq x_n$, and is minimal when either $x_1 = 0$ or $0 < x_1 \leq x_2 = x_3 = ... = x_n$.

(b) For $q \in]-\infty, 0] \cup [1, p]$, E is minimal when $0 \leq x_1 = x_2 = ... = x_{n-1} \leq x_n$, and is maximal when either $x_1 = 0$ or $0 < x_1 \leq x_2 = x_3 = ... = x_n$.

1.1.3 Useful inequalities

Theorem 39. *Let a, b, c be real numbers such that $a + b + c = s$. Let $ab + bc + ca = \frac{s^2 - t^2}{3}$ $(t \geq 0)$ and $abc = r$. Then:*

$$\frac{(s+t)^2(s-2t)}{27} \leq r \leq \frac{(s-t)^2(s+2t)}{27}$$

Equality occurs if and only if $(a - b)(b - c)(c - a) = 0$

Theorem 40 (Pham Huu Duc). *Let a, b, c, x, y, z be non-negative real numbers. Then, we have:*

$$x(a + b) + y(b + c) + z(c + a) \geq 2\sqrt{(xy + yz + zx)(ab + bc + ca)}$$

Theorem 41 (Nesbitt). *Let a, b, c be positive real numbers. Then, we have:*

$$\frac{a}{b+c} + \frac{b}{c+a} + \frac{c}{a+b} \geq \frac{3}{2}$$

Theorem 42 (Belabess). *Let $a_1, a_2, ..., a_n$ and $b_1, b_2, ..., b_n$ be non-negative real numbers such that $\max_{1 \leq i \leq n} a_i \leq \min_{1 \leq i \leq n} b_i$. The following inequality holds for all non-negative real numbers $w_1, w_2, ..., w_n$:*

$$\sum_{i=1}^{n} w_i(a_i + b_i) \geq 2\sqrt{\left(\sum_{i=1}^{n} w_i\right)\left(\sum_{i=1}^{n} w_i a_i b_i\right)}$$

Theorem 43. *Let x, y, z be real numbers such that $x + y + z \geq 0$ and $xy + yz + zx \geq 0$. The following inequality holds for all real numbers a, b, c:*

$$(b - c)^2 x + (c - a)^2 y + (a - b)^2 z \geq 0$$

Theorem 44. *For all non-negative real numbers a, b, c, we have:*

$$a^2 + b^2 + c^2 + 2abc + 1 \geq 2(ab + bc + ca)$$

with equality at $a = b = c = 1$.

Theorem 45. *Let a, b, c be non-negative real numbers. Then, we have:*

$$(a + b)(b + c)(c + a) \geq \frac{8}{9}(a + b + c)(ab + bc + ca)$$

Theorem 46. *Let a, b, c be real numbers and let $p = a + b + c$ and $q = ab + bc + ca$. Then, we have:*

$$|9(a^2 b + b^2 c + c^2 a) - p^3| \leq 2(p^2 - 3q)^{\frac{3}{2}}$$

Theorem 47 (Iran 1996). *Let a, b, c be positive real numbers. Then, we have:*

$$(ab + bc + ca)\left(\frac{1}{(a+b)^2} + \frac{1}{(b+c)^2} + \frac{1}{(c+a)^2}\right) \geq \frac{9}{4}$$

Theorem 48. *Let a, b, c be non-negative real numbers. Then, we have:*

$$a^2 b + b^2 c + c^2 a + abc \leq \frac{4(a + b + c)^3}{27}$$

Theorem 49. *Let a, b, c be positive real numbers. Then, we have:*

$$\frac{a}{b} + \frac{b}{c} + \frac{c}{a} \geq \frac{\sqrt{3(a^2 + b^2 + c^2)}}{\sqrt[3]{abc}}$$

Theorem 50. *Let a, b, c be positive real numbers. Then, we have:*

$$\frac{a}{a+b} + \frac{b}{b+c} + \frac{c}{c+a} \geq \frac{a+b+c}{a+b+c - \sqrt[3]{abc}}$$

Theorem 51. *Let a, b, c be non-negative real numbers. Then, we have:*

$$\sum_{cyc} \sqrt{(a+b)(a+c)} \geq a+b+c + \sqrt{3} \cdot \sqrt{ab + bc + ca}$$

Theorem 52 (Walker-Tooren). *Let a, b, c be positive real numbers. Then, we have:*

$$(abc)^2 (a+b+c)^3 \prod_{cyc}(b+c-a) \geq (a^2 + b^2 + c^2)^3 \prod_{cyc}(b^2 + c^2 - a^2)$$

Theorem 53. *For any two non-negative real numbers a and b, we have:*

$$\frac{1}{(1+a)^2} + \frac{1}{(1+b)^2} \geq \frac{1}{1+ab}$$

Theorem 54 (Cesaro). *Let a, b, c be non-negative real numbers. Then, we have:*

$$(a+b)(b+c)(c+a) \geq 8abc$$

Theorem 55 (Mildorf). *Let $k \geq -1$ be an integer. For any two positive real numbers a and b, we have:*

$$\frac{(1+k)(a-b)^2 + 8ab}{4(a+b)} \geq \sqrt[k]{\frac{a^k + b^k}{2}}$$

with equality if and only if $a = b$ or $k \in \{-1, 1\}$, where the power mean $k = 0$ is interpreted to be the geometric mean $\sqrt{ab}$. Moreover, if $k < -1$, then the inequality holds in the reverse direction, with equality if and only if $a = b$.

1.2   Geometric inequalities

Theorem 56 (Triangle inequality). *For any triangle, the sum of the lengths of any two sides is greater than the length of the remaining side.*

Theorem 57 (Isoperimetric inequality). *Let A be the area enclosed by a curve C of length L. We have:*

$$L^2 \geq 4\pi A$$

Equality holds if and only if C is a circle.

Theorem 58 (Euler). *Let R and r be the radii of the circumcircle and incircle of the triangle $\triangle ABC$. Then, we have:*

$$R \geq 2r$$

The equality holds if and only if the triangle $\triangle ABC$ is equilateral.

Theorem 59 (Blundon). *For any triangle with circumradius R, inradius r, and semiperimeter s, it is true that:*

$$|s^2 - (2R^2 + 10Rr - r^2)| \leq 2(R - 2r)\sqrt{R(R - 2r)}$$

Theorem 60 (Blundon). *For any triangle with circumradius R, inradius r, and semiperimeter s, it is true that:*

$$s \leq 2R + (3\sqrt{3} - 4)r$$

The equality occurs if and only if the triangle $\triangle ABC$ is equilateral.

Theorem 61 (Gerretsen). *For any triangle with circumradius R, inradius r, and semiperimeter s, it is true that:*

$$16Rr - 5r^2 \leq s^2 \leq 4R^2 + 4Rr + 3r^2$$

Theorem 62 (Ciamberlini). *Let R, r, s be the circumradius, inradius and semiperimeter respectively of an acute-angled triangle. Then:*

$$s \geq 2R + r$$

holds true; the inequality is reverse for any non-acute triangle. The equality occurs if and only if the triangle is a right triangle.

Theorem 63 (Walker). *Let R, r, s be the circumradius, inradius and semiperimeter respectively of an acute-angled triangle. Then:*

$$s^2 \geq 2R^2 + 8Rr + 3r^2$$

Theorem 64 (Hayashi). *For any triangle $\triangle ABC$ and for any arbitrary point P, we have:*

$$a.PB.PC + b.PC.PA + c.PA.PB \geq abc$$

with equality holding if and only if P is the orthocentre or one of the vertices of the triangle $\triangle ABC$.

Theorem 65 (Sandor). *For a triangle $\triangle ABC$ with area S and for any arbitrary point P, the following inequality holds:*

$$(PA.PB)^2 + (PB.PC)^2 + (PC.PA)^2 \geq \frac{16}{9}S^2$$

Theorem 66 (Erdos-Mordell). *From a point P inside a given triangle $\triangle ABC$ the perpendiculars PL,PM,PN are drawn to its sides. Then:*

$$PA + PB + PC \geq 2(PL + PM + PN)$$

Equality holds only for the equilateral triangle, where P is its centroid.

Theorem 67 (Barrow). *Let P be an arbitrary point inside the triangle $\triangle ABC$. From P and ABC, define U, V, and W as the points where the angle bisectors of BPC, CPA, and APB intersect the sides BC, CA, AB, respectively. Then:*

$$PA + PB + PC \geq 2(PU + PV + PW)$$

Equality holds only in the case of an equilateral triangle.

Theorem 68 (Weitzenböck). *For a triangle of side lengths a, b, c, and area S, the following inequality holds:*

$$a^2 + b^2 + c^2 \geq 4\sqrt{3}S$$

Theorem 69 (Neuberg-Pedoe). *If a, b, and c are the lengths of the sides of a triangle with area S_1, and A, B, and C are the lengths of the sides of a triangle with area S_2, then:*

$$A^2(b^2 + c^2 - a^2) + B^2(c^2 + a^2 - b^2) + C^2(a^2 + b^2 - c^2) \geq 16S_1 S_2$$

with equality if and only if the two triangles are similar with pairs of corresponding sides (A, a), (B, b), and (C, c).

Theorem 70 (Hadwiger–Finsler). *Let a, b, c be the side lengths of a triangle with area S. Then:*

$$a^2 + b^2 + c^2 \geq (a - b)^2 + (b - c)^2 + (c - a)^2 + 4\sqrt{3}S$$

Theorem 71 (Carlitz). *For a triangle of side lengths a, b, c, and area S, the following inequality holds:*

$$(abc)^2 \geq \left(\frac{4S}{\sqrt{3}}\right)^3$$

Theorem 72 ((Tsintsifa). *Let p, q, r be positive real numbers and a, b, c are the side lengths of a triangle with area S. Then:*

$$\frac{p}{q+r}a^2 + \frac{q}{r+p}b^2 + \frac{r}{p+q}c^2 \geq 2\sqrt{3}S$$

Theorem 73 (Ono). *Let a, b, c be the side lengths of an acute triangle with area S. Then:*

$$27(b^2 + c^2 - a^2)^2(c^2 + a^2 - b^2)^2(a^2 + b^2 - c^2)^2 \leq (4S)^6$$

Theorem 74 (Padoa). *Let a, b, c be the side lengths of a triangle. Then:*

$$abc \geq (a + b - c)(b + c - a)(c + a - b)$$

Theorem 75 (Ptolemy). *For any points A, B, C, D in the plane, we have:*

$$AB.CD + BC.DA \geq AC.BD$$

with equality if and only if $ABCD$ is a cyclic quadrilateral.

Theorem 76 (Leuenberger). *Let m_a, m_b, m_c be the medians of a triangle $\triangle ABC$. If we note by R and r the circumradius and inradius, then:*

$$m_a + m_b + m_c \leq 4R + r$$

Theorem 77 (Mitrinovic). *In any triangle with semi-perimeter s and inradius r, we have:*

$$s \geq 3\sqrt{3}r$$

Equality holds if and only if the triangle is equilateral.

Theorem 78 (Klamkin). *For any real numbers x, y, z, integer n and angles α, β, γ of any triangle, we have:*

$$x^2 + y^2 + z^2 \geq (-1)^{n+1}2(yz \cos n\alpha + zx \cos n\beta + xy \cos n\gamma)$$

Equality holds if and only if:

$$\frac{x}{\cos n\alpha} = \frac{y}{\cos n\beta} = \frac{z}{\cos n\gamma}$$

Theorem 79 (Kooi). *Let $\triangle ABC$ be a triangle with side lengths a, b, c and circumradius R. For real numbers x, y, z with $x + y + z \neq 0$, we have:*

$$(x + y + z)^2 R^2 \geq yza^2 + zxb^2 + xyc^2$$

Theorem 80 (Oppenheim). *Let x, y, z be positive real numbers and $\triangle ABC$ a triangle. S_{ABC} denotes the area and a, b, c the side lengths of the triangle. Then:*

$$a^2 x + b^2 y + c^2 z \geq 4S_{ABC}\sqrt{xy + yz + zx}$$

Theorem 81 (Birsan). *Let R, r be the circumradius and inradius respectively, d the distance between the circumcenter and incenter and a the length of a side of the triangle. We have the following inequality:*

$$4((R - d)^2 - r^2) \leq a^2 \leq 4((R + d)^2 - r^2)$$

Theorem 82 (Janous). *If we note by m_a, m_b, m_c and s the three medians and semi-perimeter of a triangle, then:*

$$\frac{1}{m_a} + \frac{1}{m_b} + \frac{1}{m_c} > \frac{5}{s}$$

Theorem 83 (Abi-Khuzam). *Let A, B, C be the vertex angles of a triangle. The following inequality holds:*

$$\sin A. \sin B. \sin C \leq \left(\frac{3\sqrt{3}}{2\pi} \right)^3 ABC$$

The maximum is reached for an equilateral triangle (and therefore at $A = B = C = \frac{\pi}{3}$).

Theorem 84 (Tereshin). *In any triangle $\triangle ABC$, the inequality*

$$m_a \geq \frac{b^2 + c^2}{4R}$$

holds. Equality holds if and only if $b = c$ or the angle from A is right.

Theorem 85 (Tereshin-Belabess). *In any triangle $\triangle ABC$, the following inequality holds:*

$$\frac{b^2 + c^2}{4R} \leq m_a \leq \frac{(b+c)^2}{16r}$$

Chapter 2

3-variable inequalities

2.1 Symmetric inequalities

Problem 1. *Let* $a, b, c \geq 0$ *such that* $a + b + c = 3$. *Prove that:*

$$\frac{a^3}{(1+a)(1+b)} + \frac{b^3}{(1+b)(1+c)} + \frac{c^3}{(1+c)(1+a)} \geq \frac{a^2 + b^2 + c^2}{4}$$

Problem 2. *Let* a, b, c *be non-negative real numbers. Prove that:*

$$4(a^2 + b^2 + c^2)^3 \geq 3(a^3 + b^3 + c^3 + 3abc)^2$$

Problem 3. *Let* $a, b, c > 0$ *and* $a^3 + b^3 + c^3 + 3abc = 6$. *Prove that:*

$$\frac{a^2 + b^2}{a + b} + \frac{b^2 + c^2}{b + c} + \frac{c^2 + a^2}{c + a} \geq 3$$

Problem 4. *Let* $a, b, c > 0$ *and* $a^3 + b^3 + c^3 + 3abc = 6$. *Prove that:*

$$\frac{a(a^2 + bc)}{b + c} + \frac{b(b^2 + ca)}{c + a} + \frac{c(c^2 + ab)}{a + b} \geq 3$$

Problem 5. *Let* $a, b, c > 0$ *such that* $ab + bc + ca = 3$. *Prove that:*

$$\frac{a^3 + b^3 + c^3}{a^2 + b^2 + c^2} + 1 \geq \frac{2}{3}(a + b + c)$$

Problem 6. *Let* $a, b, c \geq 0$ *and* $a^3 + b^3 + c^3 + 3abc = 6$. *Prove that:*

$$5(a + b + c) \geq 9 + 6abc$$

Problem 7. *Let* $a, b, c \geq 0$ *such that* $a + b + c = 3$. *Prove that:*

$$\frac{a^2 + b^2}{1 + ab} + \frac{b^2 + c^2}{1 + bc} + \frac{c^2 + a^2}{1 + ca} \geq \sqrt{3(a^2 + b^2 + c^2)}$$

Problem 8. *Let* $a, b, c > 0$ *such that* $a^3 + b^3 + c^3 = 3$. *Prove that:*

$$\frac{1}{a + b} + \frac{1}{b + c} + \frac{1}{c + a} + \frac{abc}{3} \geq \frac{11}{6}$$

Problem 9. *Let* $a, b, c \geq 0$ *real numbers such that* $ab + bc + ca + abc = 4$. *Prove that:*

$$\frac{a-1}{b+c} + \frac{b-1}{c+a} + \frac{c-1}{a+b} \geq \frac{3(a+b+c-3)}{4}$$

Problem 10. *Let* a, b, c *be positive real numbers. Prove that:*

$$\frac{(a+b)^2}{c} + \frac{(b+c)^2}{a} + \frac{(c+a)^2}{b} \geq 12\sqrt{\frac{a^3+b^3+c^3}{a+b+c}}$$

Problem 11. *Let* a, b, c *be positive real numbers satisfying* $a + b + c = 3$. *Prove that:*

$$\frac{1}{a+b} + \frac{1}{b+c} + \frac{1}{c+a} + \frac{2(a-1)(b-1)(c-1)}{3} \geq \frac{3}{2}$$

Problem 12. *Let* $a, b, c \geq 0$ *such that* $ab + bc + ca + abc = 4$. *Prove that:*

$$a^2 + b^2 + c^2 \geq 3 + 2(\sqrt{2}+1)(a+b+c-3)$$

Problem 13. *Let* $a, b, c \geq 0$ *such that* $ab + bc + ca + abc = 4$. *Prove that:*

$$\frac{(a+4)(a-1)}{a+1} + \frac{(b+4)(b-1)}{b+1} + \frac{(c+4)(c-1)}{c+1} \geq 0$$

Problem 14. *Let* a, b, c *be positive real numbers such that* $abc = 1$. *Prove that:*

$$\frac{a^3+1}{b^2+c^2} + \frac{b^3+1}{c^2+a^2} + \frac{c^3+1}{a^2+b^2} \geq a+b+c$$

Problem 15. *Let* $a, b, c \geq 0$ *such that* $ab + bc + ca = a + b + c$. *Prove that:*

$$a^3 + b^3 + c^3 + 8(a+b+c) \geq 4(a^2+b^2+c^2) + 15abc$$

Problem 16. *Let* a, b, c *be positive real numbers such that* $a + b + c = abc$. *Prove that:*

$$(4a^2 - 3)(b-c)^2 + (4b^2 - 3)(c-a)^2 + (4c^2 - 3)(a-b)^2 \geq 0$$

Problem 17. *Let* $a, b, c \geq 0$ *and* $a + b + c = 3$. *Prove that:*

$$a^2(b-c)^2 + b^2(c-a)^2 + c^2(a-b)^2 \geq \frac{9abc(1-abc)}{2}$$

Problem 18. *Let* $a, b, c \geq 0$ *such that* $a^2 + b^2 + c^2 = 3$. *Prove that:*

$$3(a^3 + b^3 + c^3) + 2abc \geq 11$$

Problem 19. *Let* $a, b, c \geq 0$ *such that* $a^2 + b^2 + c^2 = 3$. *Prove that:*

$$3 \leq \sqrt{a^2 - ab + b^2} + \sqrt{b^2 - bc + c^2} + \sqrt{c^2 - ca + a^2} \leq 3\sqrt{\frac{3}{2}}$$

Problem 20. *Let* a, b, c *be positive real numbers. Prove that:*

$$\frac{1}{\sqrt[3]{a^3+b^3}} + \frac{1}{\sqrt[3]{b^3+c^3}} + \frac{1}{\sqrt[3]{c^3+a^3}} \geq \frac{3}{\sqrt[3]{2}} \frac{a+b+c}{a^2+b^2+c^2}$$

Problem 21. *Let a, b, c be non-negative real numbers such that $a^2 + b^2 + c^2 = 3$. Prove that:*

$$(a + 2)(b + 2)(c + 2) \geq 3(a + b + c)^2$$

Problem 22. *Let a, b, c be non-negative real numbers such that $a + b + c = 3$. Prove that:*

$$a(a - b)(a - c) + b(b - a)(b - c) + c(c - a)(c - b) \geq \frac{(a - b)^2(b - c)^2(c - a)^2}{2}$$

Problem 23. *Let a, b, c be positive real numbers such that $a^2 + b^2 + c^2 = 3$. Prove that:*

$$\frac{a^3}{a^2 + bc} + \frac{b^3}{b^2 + ca} + \frac{c^3}{c^2 + ab} \geq \frac{3}{2}$$

Problem 24. *Let a, b, c be non negative real numbers. Prove that:*

$$\sqrt{1 + a^2 + b^2 + c^2 - ab - bc - ca} \geq \frac{a + b + c - abc}{2}$$

Problem 25. *Let $a, b, c \geq 0$ such that $a + b + c = ab + bc + ca$. Prove that:*

$$\sqrt{1 + (a - b)^2} + \sqrt{1 + (b - c)^2} + \sqrt{1 + (c - a)^2} \geq a + b + c$$

Problem 26. *Let a, b, c be positive real numbers. Prove that:*

$$\frac{(a + b - c)^2}{(a + b)^2 + c^2} + \frac{(b + c - a)^2}{(b + c)^2 + a^2} + \frac{(c + a - b)^2}{(c + a)^2 + b^2} \geq \frac{9(a^2 + b^2 + c^2)}{5(a + b + c)^2}$$

Problem 27. *Let $a, b, c > 0$ and $a + b + c = 3$. Prove that:*

$$\left(\frac{a}{b} + \frac{b}{c} + \frac{c}{a} \right) \left(\frac{b}{a} + \frac{c}{b} + \frac{a}{c} \right) \geq 2(a^3 + b^3 + c^3) + 3$$

Problem 28. *Let a, b, c be positive real numbers. Prove that:*

$$\sqrt[3]{\frac{a^2}{b^2 + c^2}} + \sqrt[3]{\frac{b^2}{c^2 + a^2}} + \sqrt[3]{\frac{c^2}{a^2 + b^2}} \geq \frac{\sqrt[3]{3}(a + b + c)}{\sqrt[3]{2(a^3 + b^3 + c^3)}}$$

Problem 29. *Let a, b, c be positive real numbers. Prove that:*

$$\frac{a}{b + c} + \frac{b}{c + a} + \frac{c}{a + b} \geq \frac{3}{2} + \frac{4(a^4 + b^4 + c^4 - abc(a + b + c))}{(a + b + c)^4}$$

Problem 30. *Let a, b, c be positive real numbers such that $a + b + c = 3$. Prove that:*

$$\frac{ab + c}{a + b} + \frac{bc + a}{b + c} + \frac{ca + b}{c + a} \geq \frac{11}{4}$$

Problem 31. *Let a, b, c be positive real numbers. Prove that:*

$$\frac{\sum_{cyc} a^2}{\sum_{cyc} ab} + \frac{1}{2} \sum_{cyc} \frac{ab}{a^2 + b^2} \geq \frac{7}{4}$$

17

Problem 32. *Let a, b, c be positive real numbers such that $a + b + c = 3$. Prove that:*

$$\frac{1}{(a+b)^2} + \frac{1}{(b+c)^2} + \frac{1}{(c+a)^2} + \frac{abc}{4} \geq 1$$

Problem 33. *For any positive real numbers a, b and c, prove that:*

$$\sum_{cyc} \frac{a^4 + b^4}{a^2 + b^2} \geq \frac{2(a^3 + b^3 + c^3) + 3abc}{a + b + c}$$

Problem 34. *Let $a, b, c > 0$ such that $a + b + c = \frac{1}{a} + \frac{1}{b} + \frac{1}{c}$. Prove that:*

$$\sqrt{1 + (a - b)^2} + \sqrt{1 + (b - c)^2} + \sqrt{1 + (c - a)^2} \geq a + b + c$$

Problem 35. *Let a, b, c be non-negative real numbers such that $a^3 + b^3 + c^3 = 3$. Prove that:*

$$\sqrt{3(a^2 + b^2 + c^2)} \geq 1 + \frac{2(a + b + c)}{3}$$

Problem 36. *Let a, b, c be positive real numbers. Prove that:*

$$\frac{a^3}{(b+c)^2} + \frac{b^3}{(c+a)^2} + \frac{c^3}{(a+b)^2} \geq \frac{11\left(a^2 + b^2 + c^2\right) - 8\left(ab + bc + ca\right)}{4(a + b + c)}$$

Problem 37. *Let a, b, c be non-negative real numbers such that $a^2 + b^2 + c^2 = 3$. Prove that:*

$$\sqrt[3]{\frac{a^3 + b^3 + c^3}{3}} + \frac{a + b + c}{9} \geq \frac{4}{3}$$

Problem 38. *Let $a, b, c > 0$ such that $a^2 + b^2 + c^2 + abc = 4$. Prove that:*

$$\frac{1}{a^2 + b^2} + \frac{1}{b^2 + c^2} + \frac{1}{c^2 + a^2} \geq \frac{5}{4}$$

Problem 39. *Let a, b, c be positive real numbers. Prove that:*

$$\frac{a^2 + bc}{b^2 + c^2} + \frac{b^2 + ca}{c^2 + a^2} + \frac{c^2 + ab}{a^2 + b^2} \geq \frac{5}{2}$$

Problem 40. *Let a, b, c be positive real numbers. Prove that:*

$$\frac{a(a^2 + bc)}{a + b} + \frac{b(b^2 + ca)}{b + c} + \frac{c(c^2 + ab)}{c + a} \geq \frac{(a + b + c)^2}{3}$$

Problem 41. *Let a, b, c be non-negative real numbers such that $a + b + c = 3$. Prove that:*

$$(5a^2 + 6)(5b^2 + 6)(5c^2 + 6) \geq 1331$$

Problem 42. *Let $a, b, c \geq 0$ such that $a^2 + b^2 + c^2 = 3$. Prove that:*

$$a^4 b^4 + b^4 c^4 + c^4 a^4 + (abc)^3 \geq 4(abc)^2$$

Problem 43. *Let $a, b, c > 0$ such that $a^3 + b^3 + c^3 + 3abc = 6$. Prove that:*

$$\frac{1}{a^3} + \frac{1}{b^3} + \frac{1}{c^3} + 3 \geq 2(a^3 + b^3 + c^3)$$

Problem 44. *Let $a, b, c > 0$ such that $a + b + c = ab + bc + ca$. Prove that:*

$$a + b + c - \frac{3}{2} \geq \frac{a}{b + c} + \frac{b}{c + a} + \frac{c}{a + b} \geq \frac{a + b + c}{2}$$

Problem 45. *Let $a, b, c \geq 0$ such that $ab + bc + ca + abc = 4$. Prove that:*

$$\sqrt{ab} + \sqrt{bc} + \sqrt{ca} \leq 3$$

Problem 46. *Let $a, b, c > 0$ and $a^3 + b^3 + c^3 + 3abc = 6$. Prove that:*

$$\frac{1}{a^2 + 2bc} + \frac{1}{b^2 + 2ca} + \frac{1}{c^2 + 2ab} \geq \frac{a^2 + b^2 + c^2 + 3}{6}$$

Problem 47. *Let a, b, c be positive real numbers such that $a + b + c = 3$. Prove that:*

$$2^{a+1} + 2^{b+1} + 2^{c+1} + abc \geq 13$$

Problem 48. *Let a, b, c be non-negative real numbers such that $abc(a+b+c) = 3$. Prove that:*

$$12 \sum_{cyc} (a^2 - ab) \leq \prod_{cyc} (a + b) \sum_{cyc} c(a - b)^2$$

Problem 49. *Let a, b, c be positive real numbers such that $a + b + c = 1$. Prove that:*

$$\frac{a + b}{(a - b)^2} + \frac{b + c}{(b - c)^2} + \frac{c + a}{(c - a)^2} \geq 9$$

Problem 50. *Let $a, b, c \geq 0$ such that $abc = 1$. Prove that:*

$$(a^3 + 1)(b^3 + 1)(c^3 + 1) \geq 4(a\sqrt{a} + b\sqrt{b} + c\sqrt{c} - 1)$$

Problem 51. *Let a, b, c be positive numbers such that $ab + bc + ca = 3$. Prove that:*

$$\sqrt{a + b} + \sqrt{b + c} + \sqrt{c + a} \geq 3\sqrt{2}$$

Problem 52. *Let a, b, c be positive real numbers such that $abc = 1$. Prove that:*

$$(a + b)^{a+b} + (b + c)^{b+c} + (c + a)^{c+a} \geq 4(a + b + c)$$

Problem 53. *Let a, b, c be positive real numbers such that $a + b + c = 3$. Prove that:*

$$\frac{a^3 + b^3}{1 + ab} + \frac{b^3 + c^3}{1 + bc} + \frac{c^3 + a^3}{1 + ca} \geq \sqrt{3(a^3 + b^3 + c^3)}$$

2.2 Non-symmetric inequalities

Problem 54. *Let a, b, c be real numbers such that $a^2 + b^2 + c^2 = 3$. Prove that:*

$$a(2a^2 + 3b^2) + b(2b^2 + 3c^2) + c(2c^2 + 3a^2) \leq 15$$

Problem 55. *Let a, b, c be non-negative real numbers such that $a + b + c = 3$. Prove that:*

$$\frac{a}{b^2 + 1} + \frac{b}{c^2 + 1} + \frac{c}{a^2 + 1} \geq \frac{\sqrt{3(a^2 + b^2 + c^2)}}{2}$$

Problem 56. *Let* $a, b, c > 0$ *such that* $a^2 + b^2 + c^2 = 3$. *Prove that:*

$$\frac{a}{b} + \frac{b}{c} + \frac{c}{a} + abc \geq 4$$

Problem 57. *Let* a, b, c *be non-negative real numbers. Prove that:*

$$a^3 + b^3 + c^3 + 4(a - b)(b - c)(c - a) \geq 3abc$$

Problem 58. *Let* a, b, c *be positive real numbers such that* $\frac{a}{c} + \frac{b}{c} + \frac{c}{a} \geq \frac{b}{a} + \frac{c}{b} + \frac{a}{c}$. *Prove that:*

$$\frac{a}{b} + \frac{b}{c} + \frac{c}{a} + 1 \geq \frac{12(a^2 + b^2 + c^2)}{(a + b + c)^2}$$

Problem 59. *Let* a, b, c *be positive real numbers. Prove that:*

$$\frac{a}{b} + \frac{b}{c} + \frac{c}{a} \geq \frac{a}{b + c} + \frac{b}{c + a} + \frac{c}{a + b} + \frac{3}{2}$$

Problem 60. *Let* $a, b, c > 0$ *such that* $abc = 1$. *Prove that:*

$$5\left(\frac{a}{b} + \frac{b}{c} + \frac{c}{a}\right) + 12 \geq 9(ab + bc + ca)$$

Problem 61. *Let* $a, b, c \geq 0$ *such that* $a^2 + b^2 + c^2 = 3$. *Prove that:*

$$\frac{a}{b + 1} + \frac{b}{c + 1} + \frac{c}{a + 1} \geq \frac{3}{2}$$

Problem 62. *Let* $a, b, c \geq 0$ *such that* $a^2 + b^2 + c^2 = 3$. *Prove that:*

$$\frac{a}{b^2 + 1} + \frac{b}{c^2 + 1} + \frac{c}{a^2 + 1} \geq \frac{3}{2}$$

Problem 63. *Let* a, b, c *be positive real numbers such that* $abc = 1$. *Prove that:*

$$(a^2 + b^2)(b^2 + c^2)(c^2 + a^2) \geq 4\left(\frac{a}{b} + \frac{b}{c} + \frac{c}{a} - 1\right)$$

Problem 64. *Let* a, b, c *be non-negative real numbers such that* $a + b + c = abc$. *Prove that:*

$$\frac{a + 1}{b^2 + 1} + \frac{b + 1}{c^2 + 1} + \frac{c + 1}{a^2 + 1} \geq \frac{3(\sqrt{3} + 1)}{4}$$

Problem 65. *Let* a, b, c *be positive real numbers. Prove that:*

$$5\left(\frac{a}{b} + \frac{b}{c} + \frac{c}{a}\right)^2 \geq 2(a + b + c)\left(\frac{1}{a} + \frac{1}{b} + \frac{1}{c}\right) + 27$$

Problem 66. *Let* a, b, c *be positive real numbers such that* $a + b + c = 3$. *Prove that:*

$$\frac{a + b}{b + c} + \frac{b + c}{c + a} + \frac{c + a}{a + b} \geq \frac{a^2 + b^2 + c^2 + 9}{4}$$

Problem 67. *Let* $a, b, c > 0$ *such that* $a + b + c = \frac{1}{a} + \frac{1}{b} + \frac{1}{c}$. *Prove that:*

$$\frac{a}{b} + \frac{b}{c} + \frac{c}{a} + 3 \geq 2(a + b + c)$$

Problem 68. *Let* a, b, c *be positive real numbers such that* $a + b + c = 3$. *Prove that:*

$$\frac{a-b}{b} + \frac{b-c}{c} + \frac{c-a}{a} \geq 2(a-b)(b-c)(c-a)$$

Problem 69. *Let* a, b, c *be non-negative real numbers such that* $a + b + c = 1$. *Prove that:*

$$2\sqrt{3} \geq \sqrt{a}(b + 2c + 1) + \sqrt{b}(c + 2a + 1) + \sqrt{c}(a + 2b + 1) \geq 1.$$

Problem 70. *Let* a, b, c *be positive real numbers such that* $a + b + c = 1$. *Find the maximum and the minimum values of the following expression:*

$$\sqrt{a}(b + 2c - 1) + \sqrt{b}(c + 2a - 1) + \sqrt{c}(a + 2b - 1)$$

Problem 71. *Let* a, b, c *be positive numbers. Prove that:*

$$\frac{a}{a+b} + \frac{b}{b+c} + \frac{c}{c+a} + \frac{a+b+c}{4\sqrt[3]{abc}} \geq \frac{9}{4}$$

Problem 72. *Let* a, b, c *be non-negative real numbers. Prove that:*

$$(a^3 + b^3 + c^3)^5 \geq 9(a^4b + b^4c + c^4a)^3$$

Problem 73. *Let* $a, b, c > 0$ *such that* $abc = 1$. *Prove that:*

$$\frac{a^2+1}{a+b} + \frac{b^2+1}{b+c} + \frac{c^2+1}{c+a} \geq \sqrt{3(a+b+c)}$$

Problem 74. *Let* a, b, c *be positive real numbers such that* $abc = 1$. *Prove that:*

$$(a^2 + b^2)(b^2 + c^2)(c^2 + a^2) \geq 4\left(\frac{a}{b} + \frac{b}{c} + \frac{c}{a} - 1\right)$$

Problem 75. *Let* a, b, c *be positive real numbers such that* $abc = 1$. *Prove that:*

$$\frac{a^2+a+1}{b} + \frac{b^2+b+1}{c} + \frac{c^2+c+1}{a} \geq 3\sqrt{3(a^2 + b^2 + c^2)}$$

Problem 76. *Let* a, b, c *be positive real numbers with* $a + b + c = 3$. *Prove that:*

$$\frac{a}{b} + \frac{b}{c} + \frac{c}{a} \geq a^2 + b^2 + c^2$$

Problem 77. *Let* a, b, c *be positive real numbers such that* $ab + bc + ca = 3$. *Prove that:*

$$\sqrt{7a^2 + b + \frac{1}{c}} + \sqrt{7b^2 + c + \frac{1}{a}} + \sqrt{7c^2 + a + \frac{1}{b}} \geq 9$$

2.3 Challenges

Challenge 1. *Let* $a, b, c \geq 0$ *and* $a^3 + b^3 + c^3 + 3abc = 6$. *Prove that:*

$$\frac{a^2+1}{a+1} + \frac{b^2+1}{b+1} + \frac{c^2+1}{c+1} \geq 3$$

Challenge 2. *Let* $a, b, c \geq 0$ *and* $a^3 + b^3 + c^3 + 3abc = 6$. *Prove that:*

$$a^2(b-c)^2 + b^2(c-a)^2 + c^2(a-b)^2 \geq \frac{5abc(1-abc)}{2}$$

Challenge 3. *Let* $a, b, c > 0$ *such that* $a^3 + b^3 + c^3 + 3abc = 6$. *Prove that:*

$$\frac{4a^2 + 1}{a} + \frac{4b^2 + 1}{b} + \frac{4c^2 + 1}{c} \geq 15$$

Challenge 4. *Let* $a, b, c > 0$ *such that* $a^3 + b^3 + c^3 + 3abc = 6$. *Prove that:*

$$\frac{1}{a^2} + \frac{1}{b^2} + \frac{1}{c^2} + 15 \geq 6(a^2 + b^2 + c^2)$$

Challenge 5. : *Let* a, b, c *be positive real numbers satisfying* $a + b + c = 3$. *Prove that:*

$$\frac{a^2 - bc}{b^2 + c^2} + \frac{b^2 - ca}{c^2 + a^2} + \frac{c^2 - ab}{a^2 + b^2} \geq \frac{3(1 - abc)}{2}$$

Challenge 6. *Let* $a, b, c \geq 0$ *such that* $a^2 + b^2 + c^2 = 3$. *Find the maximum value of:*

$$\sqrt{a + b} + \sqrt{b + c} + \sqrt{c + a} - \frac{abc}{3}$$

Challenge 7. *For any positive real numbers* a, b, c *such that* $a + b + c = 3$. *Prove that:*

$$(9 - 5a^2)(9 - 5b^2)(9 - 5c^2) \leq 64$$

Challenge 8. *Let* $a, b, c \geq 0$ *such that* $a^2 + b^2 + c^2 = 3$. *Prove that:*

$$\frac{a}{b^2 + 1} + \frac{b}{c^2 + 1} + \frac{c}{a^2 + 1} + \frac{abc}{5} \geq \frac{17}{10}$$

Challenge 9. : *Let* a, b, c *be positive real numbers such that* $a^2 + b^2 + c^2 = 3$. *Prove that:*

$$\frac{a^2}{b^2 + 2ab} + \frac{b^2}{c^2 + 2bc} + \frac{c^2}{a^2 + 2ca} \geq \frac{a^3 + b^3 + c^3}{3}$$

Challenge 10. :*Let* a, b, c *be positive real numbers. Prove that:*

$$\frac{(a + 2b)^3}{(a + 2c)^3} + \frac{(b + 2c)^3}{(b + 2a)^3} + \frac{(c + 2a)^3}{(c + 2b)^3} \geq \frac{27(a^3 + b^3 + c^3)}{(a + b + c)^3}$$

Challenge 11. *Let* $a, b, c \geq 1$ *such that* $a + b + c = \frac{1}{a} + \frac{1}{b} + \frac{1}{c} + 8$

$$4(ab + bc + ca) + 5(a + b + c) \leq 153$$

Challenge 12. *Let* a, b, c *be real numbers. Prove that:*

$$2(a^4 + b^4 + c^4) \geq (a + b + c)(a^3 + b^3 + c^3 - abc)$$

Challenge 13. *Let* a, b, c *be positive real numbers such that* $a + b + c = 3$. *Prove that:*

$$\frac{ab}{c^2 + ca} + \frac{bc}{a^2 + ab} + \frac{ca}{b^2 + bc} \geq \frac{a^2 + b^2 + c^2}{2}$$

Challenge 14. *Let a, b, c be positive real numbers such that $a + b + c = 3$. Prove that:*

$$\frac{a(b+c)}{c(c+a)} + \frac{b(c+a)}{a(a+b)} + \frac{c(a+b)}{b(b+c)} \geq \frac{2(a^2 + b^2 + c^2)}{3} + 1$$

Challenge 15. *Let a, b, c be positive real numbers. Prove that:*

$$\frac{a}{b+c} + \frac{b}{c+a} + \frac{c}{a+b} \geq \frac{3}{2} + \frac{2(a^3 + b^3 + c^3 - 3abc)}{(a+b+c)^3}$$

Challenge 16. *Let $a, b, c > 0$ such that $abc = 1$. Prove that:*

$$\frac{a^2+1}{a+b} + \frac{b^2+1}{b+c} + \frac{c^2+1}{c+a} \geq \sqrt{3(a+b+c)}$$

Challenge 17. *Let a, b, c be positive real numbers such that $abc = 1$. Prove that:*

$$\sqrt[b]{a} + \sqrt[c]{b} + \sqrt[a]{c} \geq 3$$

Challenge 18. *Let a, b, c be positive real numbers such that $abc = 1$. Prove that:*

$$\sqrt[b]{1+a} + \sqrt[c]{1+b} + \sqrt[a]{1+c} \geq 6$$

Challenge 19. *Let a, b, c be positive real numbers such that $abc = 1$. Prove that:*

$$\min\left(\sqrt[b]{1+a} + \sqrt[c]{1+b} + \sqrt[a]{1+c}, \sqrt[c]{a+b} + \sqrt[a]{b+c} + \sqrt[b]{c+a}\right) \geq 2(a+b+c)$$

Challenge 20. *Let $a, b, c \geq 0$ satisfying $a + b + c = 3$. Prove that:*

$$ab(a-b)^2 + bc(b-c)^2 + ca(c-a)^2 \geq 3abc(1-abc)$$

Challenge 21. *Let a, b, c be positive real numbers such that $a + b + c = 3$. Prove that:*

$$\frac{2a}{b+c} + \frac{2b}{c+a} + \frac{2c}{a+b} + abc \geq 4$$

Challenge 22. *Let a, b, c be positive real numbers. Prove that:*

$$\frac{a^4 + b^4 + c^4}{a^2b + b^2c + c^2a} + \frac{a^2b + b^2c + c^2a}{a^2 + b^2 + c^2} \geq \frac{2(a^3b + b^3c + c^3a)}{a^2b + b^2c + c^2a}$$

Challenge 23. *Let a, b, c be positive real numbers such that $abc = 1$. Prove that:*

$$a^a + b^b + c^c + 3 \geq 2(a+b+c)$$

Challenge 24. *Let $a, b, c \geq 0$ such that $a^2 + b^2 + c^2 = 3$. Prove that:*

$$4(a+b+c) + 3abc \geq 4(ab+bc+ca) + 3$$

Challenge 25. *Let a, b, c be positive real numbers such that $abc = 1$. Prove that:*

$$\left(1 + \frac{a}{b}\right)^a \left(1 + \frac{b}{c}\right)^b \left(1 + \frac{c}{a}\right)^c \geq 2(a^2 + b^2 + c^2)$$

Challenge 26. *Let a, b, c be non-negative real numbers such that $abc = 1$. Prove that:*

$$\frac{a^2 + b^2 + 2}{\sqrt{(a^2 + 1)(b^2 + 1)}} + \frac{b^2 + c^2 + 2}{\sqrt{(b^2 + 1)(c^2 + 1)}} + \frac{c^2 + a^2 + 2}{\sqrt{(c^2 + 1)(a^2 + 1)}} \geq 2\sqrt{3(a + b + c)}$$

Challenge 27. *Let a, b, c be non-negative real numbers such that $a + b + c = 1$. Find the maximum value of:*

$$abc(a - b)(b - c)(c - a)$$

Challenge 28. *Let a, b, c be non-negative real numbers such that $a + b + c = 3$. Prove that:*

$$\frac{a^2 + b^2}{\sqrt{(a^2 + 1)(b^2 + 1)}} + \frac{b^2 + c^2}{\sqrt{(b^2 + 1)(c^2 + 1)}} + \frac{c^2 + a^2}{\sqrt{(c^2 + 1)(a^2 + 1)}} \geq \sqrt{3(a^2 + b^2 + c^2)}$$

Challenge 29. *Let a, b, c be positive real numbers such that $a + b + c = 3$. Prove that:*

$$\frac{a^3 + b^3}{1 + a} + \frac{b^3 + c^3}{1 + b} + \frac{c^3 + a^3}{1 + c} \geq a^2\sqrt{a} + b^2\sqrt{b} + c^2\sqrt{c}$$

Challenge 30. *Let a, b, c be positive real numbers such that $ab + bc + ca = 3$. Prove that:*

$$\sqrt{7a^2 + b + \frac{1}{c}} + \sqrt{7b^2 + c + \frac{1}{a}} + \sqrt{7c^2 + a + \frac{1}{b}} \geq 3\sqrt{3(a + b + c)}$$

Challenge 31. *Let $a \geq b \geq c \geq 0$ such that $ab + bc + ca = 3$. Prove that:*

$$a + b + c + (c - 1)^2 \geq 3 + (a - b)(b - c)$$

Chapter 3

4-variable inequalities

3.1 Symmetric inequalities

Problem 78. *Let* $a, b, c, d > 0$ *such that* $a + b + c + d = \frac{1}{a} + \frac{1}{b} + \frac{1}{c} + \frac{1}{d}$. *Prove that:*

$$a^2 + b^2 + c^2 + d^2 + 8 \geq 3(a + b + c + d)$$

Problem 79. *Let* $a, b, c, d \geq 0$ *such that* $a^2 + b^2 + c^2 + d^2 = 4$. *Prove that:*

$$18 + 2abcd \geq 5(a + b + c + d)$$

Problem 80. *Let* $a, b, c, d \geq 0$ *such that* $a + b + c + d = 4$. *Prove that:*

$$a^4 + b^4 + c^4 + d^4 \geq 4abcd + 32(a - 1)(b - 1)(c - 1)(d - 1)$$

Problem 81. *Let* $a, b, c, d \geq 0$ *such that* $a + b + c + d = 4$. *Prove that:*

$$\frac{1}{a^2 + 1} + \frac{1}{b^2 + 1} + \frac{1}{c^2 + 1} + \frac{1}{d^2 + 1} \geq 2$$

Problem 82. *Let* $a, b, c, d \geq 0$ *such that* $a + b + c + d + abcd = 5$. *Prove that:*

$$3(a^2 + b^2 + c^2 + d^2) + 40 \geq 13(a + b + c + d)$$

Problem 83. *Let* $a, b, c, d \geq 0$ *such that* $a + b + c + d + abcd = 5$. *Prove that:*

$$27(abc + bcd + acd + abd) + 17abcd \leq 125$$

Problem 84. *Let* a, b, c, d *be non-negative real numbers . Prove that:*

$$\sum_{cyc}(a - b)(a - c)(a - d) \geq -\frac{(a + b + c + d)^3}{27}$$

Problem 85. *Let* a, b, c, d *be non-negative real numbers . Prove that:*

$$a^2 + b^2 + c^2 + d^2 + abcd + 1 \geq ab + bc + cd + da + ac + bd$$

Problem 86. *Let* $a, b, c, d \geq 0$ *such that* $a + b + c + d + abcd = 5$. *Prove that:*

$$\frac{1}{a + 1} + \frac{1}{b + 1} + \frac{1}{c + 1} + \frac{1}{d + 1} \geq 2$$

Problem 87. *Let* $a, b, c, d \geq 0$ *such that* $a^2 + b^2 + c^2 + d^2 = 4$. *Prove that:*

$$a^3 + b^3 + c^3 + d^3 + abc + bcd + cda + dab \leq 8$$

Problem 88. *Let* $a, b, c, d > 0$ *such that* $a + b + c + d = \frac{1}{a} + \frac{1}{b} + \frac{1}{c} + \frac{1}{d}$. *Prove that:*

$$(a+1)(b+1)(c+1)(d+1) \geq 16$$

Problem 89. *Let* $a, b, c, d > 0$ *such that* $a + b + c + d = \frac{1}{a} + \frac{1}{b} + \frac{1}{c} + \frac{1}{d}$. *Prove that:*

$$a^2 + b^2 + c^2 + d^2 + 2abcd + 6 \geq 3(a + b + c + d)$$

Problem 90. *Let* $a, b, c, d > 0$ *such that* $a + b + c + d = 4$. *Prove that:*

$$\frac{a^2}{b+c+d} + \frac{b^2}{c+d+a} + \frac{c^2}{d+a+b} + \frac{d^2}{a+b+c} + \frac{2}{3} \geq \frac{1}{2}(a^2 + b^2 + c^2 + d^2)$$

Problem 91. *Let* $a, b, c, d \geq 0$ *such that* $a + b + c + d = 4$. *Prove that:*

$$\frac{a^2 + b^2 + c^2 + d^2}{4 - abcd} \geq \frac{4}{3}$$

3.2 Non-symmetric inequalities

Problem 92. *Let* a, b, c, d *be real numbers. Prove that:*

$$\sqrt{(a^2 + b^2 + 2)(b^2 + c^2 + 2)(c^2 + d^2 + 2)(d^2 + a^2 + 2)} \geq (1 + a)(1 + b)(1 + c)(1 + d)$$

Problem 93. *Let* $a, b, c, d \geq 0$ *such that* $a + b + c + d = 4$. *Prove that:*

$$a + \frac{b^2}{2} + \frac{b^3}{3} + \frac{d^4}{4} \geq \frac{25}{12}$$

Problem 94. : *Let* a, b, c, d *be non-negative real numbers. Prove that:*

$$a^4 + b^4 + c^4 + d^4 + 2(a - b)(b - c)(c - d)(d - a) \geq 4abcd$$

Problem 95. *Let* $a, b, c, d \geq 0$ *such that* $a^2 + b^2 + c^2 + d^2 = 4$. *Prove that:*

$$\frac{1}{(a - b)^2} + \frac{1}{(b - c)^2} + \frac{1}{(c - d)^2} + \frac{1}{(d - a)^2} \geq 2$$

Problem 96. *Let* $a, b, c, d \geq 0$ *and* $a + b + c + d = 4$. *Prove that:*

$$a + abc + \frac{9}{4} \geq ab + abcd$$

Problem 97. *Let* $a, b, c, d \geq 0$ *such that* $a + b + c + d = 4$. *Prove that:*

$$a^2 b + b^2 c + c^2 d + d^2 a \leq \frac{256}{27}$$

Problem 98. *Let* a, b, c, d *be positive real numbers. Prove that:*

$$\frac{a}{b} + \frac{b}{c} + \frac{c}{d} + \frac{d}{a} \geq \frac{a+b}{b+c} + \frac{b+c}{c+d} + \frac{c+d}{d+a} + \frac{d+a}{a+b}$$

Problem 99. : *Let* a, b, c, d *be real numbers such that* $a^2 + b^2 + c^2 + d^2 = 4$. *Find the min and the max of the following expression:*

$$a^2 b + b^2 c + c^2 d + d^2 a$$

3.3 Challenges

Challenge 32. *Let $a, b, c, d > 0$ such that $a + b + c + d = 4$. Prove that:*

$$\frac{a}{b} + \frac{b}{c} + \frac{c}{d} + \frac{d}{a} + 2abcd \geq 6$$

Challenge 33. *: Let a, b, c, d be positive real numbers. Prove that:*

$$\frac{a}{b} + \frac{b}{c} + \frac{c}{d} + \frac{d}{a} \geq 4 + \frac{8(a - d)^2}{(a + b + c + d)^2}$$

Challenge 34. *Let $a, b, c, d > 0$ such that $a^2 + b^2 + c^2 + d^2 = 4$. Prove that:*

$$\frac{a}{b} + \frac{b}{c} + \frac{c}{d} + \frac{d}{a} + abcd \geq 5$$

Challenge 35. *Let $a, b, c, d \geq 0$ such that $a + b + c + d = 4$. Prove that:*

$$\frac{1}{1 + ab} + \frac{1}{1 + bc} + \frac{1}{1 + cd} + \frac{1}{1 + da} + \frac{abcd}{2} \geq \frac{5}{2}$$

Challenge 36. *Let a, b, c, d be positive real numbers such that $a + b + c + d = 4$. Prove that:*

$$\frac{a - b}{b} + \frac{b - c}{c} + \frac{c - d}{d} + \frac{d - a}{a} \geq (a - b)(b - c)(c - d)(d - a)$$

Challenge 37. *Let $a, b, c, d > 0$ such that $a + b + c + d = 4$. Prove that:*

$$\frac{4 - a}{b} + \frac{4 - b}{c} + \frac{4 - c}{d} + \frac{4 - d}{a} + 4abcd \geq 16$$

Challenge 38. *Let a, b, c, d be non-negative real numbers such that $a^2 + b^2 + c^2 + d^2 = 4$. Prove that:*

$$2(a^3 + b^3 + c^3 + d^3) + abcd \geq 9$$

Challenge 39. *Let $a, b, c, d \geq 0$ such that $a + b + c + d = 4$. Prove that:*

$$\frac{1}{3 + a} + \frac{1}{3 + b} + \frac{1}{3 + c} + \frac{1}{3 + d} \leq \frac{1}{2} + \frac{2}{3 + abcd}$$

Challenge 40. *Let $a, b, c, d > 0$ such that $a + b + c + d + abcd = 5$. Prove that:*

$$\frac{1}{a^2} + \frac{1}{b^2} + \frac{1}{c^2} + \frac{1}{d^2} \geq a^2 + b^2 + c^2 + d^2$$

Challenge 41. *Let $a \geq b \geq c \geq d \geq 0$ such that $(a + c)(b + d) = 4$. Prove that:*

$$(1 - ac)\sqrt{b + d} + (1 - bd)\sqrt{a + c} \geq 0$$

Challenge 42. *Let $a, b, c, d \geq 0$ such that $a + b + c + d + abcd = 5$. Prove that:*

$$\frac{1}{a + 1} + \frac{1}{b + 1} + \frac{1}{c + 1} + \frac{1}{d + 1} \geq \frac{a + b + c + d + 12}{8}$$

Challenge 43. *Let $a, b, c, d \geq 0$ such that $a^2 + b^2 + c^2 + d^2 = 4$. Prove that:*

$$a^3 + b^3 + c^3 + d^3 + \frac{abc + bcd + cda + dab}{4} \geq 5$$

Challenge 44. *Let $a, b, c, d \geq 0$ such that $a + b + c + d + abcd = 5$. Prove that:*

$$a + b + c + d \geq abc + bcd + acd + abd$$

Challenge 45. *Let $a \leq b \leq c \leq d$ be non-negative real numbers such that any three different numbers chosen from $\{a, b, c, d\}$ constitute the side lengths of a triangle. Prove that:*

$$ad < \frac{2(a + b + c + d)^2}{25}$$

Challenge 46. *Let $a, b, c, d \geq 0$ such that $a + b + c + d = 4$. Prove that:*

$$a + \frac{b^2}{2} + \frac{b^3}{3} + \frac{d^4}{4} \geq \frac{23(a^2 + b^2 + c^2 + d^2) + 208}{144}$$

Challenge 47. *Let $a \geq b \geq c \geq d$ be non-negative real numbers such that $a + b + c + d = 1$. Prove that:*

$$(a + c)(b + d) \geq \sqrt{ac(b + d) + bd(a + c)}$$

Chapter 4

Geometric inequalities

4.1 Solved Problems

Problem 100. *Let $\triangle ABC$ be a triangle with centroid G. If we note by R, R_1, R_2 and R_3 the circumradii of triangles $\triangle ABC$, $\triangle GAB$, $\triangle GBC$ and $\triangle GCA$ respectively, prove that:*

$$R_1 + R_2 + R_3 \geq 3R$$

Problem 101. *Let $\triangle ABC$ be a triangle with side lengths a, b and c, a circumcenter O and incenter I. Prove that:*

$$0 \leq \frac{3}{2} - (\cos A + \cos B + \cos C) \leq \frac{2.OI}{a+b+c}$$

Problem 102. *Let $\triangle ABC$ be a triangle with medians of lengths m_a, m_b, m_c and R the circumradius. Prove that:*

$$\frac{m_a^2 + m_b^2 + m_c^2}{m_a + m_b + m_c} \leq \frac{3R}{2}$$

Problem 103. *Prove that in any acute triangle, the inradius of the orthic triangle cannot exceed $\frac{1}{2}$ the inradius of the original triangle.*

Problem 104. *Let $\triangle ABC$ be an acute triangle, and R, P, Q the feet of the altitudes on AB, BC and AC. If we note by r, r_1, r_2, r_3 and r_4 the inradii of the triangles $\triangle ABC$, $\triangle AQR$, $\triangle BPR$, $\triangle CPQ$, and $\triangle RPQ$, prove that:*

$$r < r_1 + r_2 + r_3 + r_4 \leq 2r$$

Problem 105. *Let R and r be the circumradius and inradius of an acute triangle $\triangle ABC$. If we note by $\alpha_H, \beta_H, \gamma_H$ the angles opposite the sides of the orthic triangle, prove that:*

$$\cos \alpha_H + \cos \beta_H + \cos \gamma_H + \frac{5R}{4r} \geq 4$$

Problem 106. *Prove that in any acute triangle, the area of the orthic triangle cannot exceed $\frac{1}{4}$ of the area of the original triangle.*

Problem 107. *Let $\triangle ABC$ be an acute triangle with side lengths a, b and c. If we note by R the circumradius, prove that:*

$$a^4 + b^4 + c^4 \geq 24R^4$$

Problem 108. *Let* m_a, m_b, m_c *be the lengths of the medians of an acute triangle and R and r are respectively the circumradius and inradius. Prove the following inequality:*

$$am_a + bm_b + cm_c \leq \frac{9\sqrt{3}R^2}{2}$$

Problem 109. *Prove that in any acute triangle, the circumradius cannot exceed $\frac{\sqrt{2}}{2}$ the largest altitude.*

Problem 110. *Let $\triangle ABC$ be an acute triangle with circumcenter O and incenter I. If we note by p_1 and p_2 the perimeters of $\triangle ABC$ and its orthic triangle respectively, prove that:*

$$0 \leq \frac{p_1}{2} - p_2 \leq 2.OI$$

Problem 111. *Let $\triangle ABC$ be an acute triangle with inradius r. If we note by p_{orthic} the perimeter of the orthic triangle, prove that:*

$$4r < p_{orthic} \leq 3\sqrt{3}r$$

Problem 112. *Let $\triangle ABC$ be a triangle and R, P, Q the feet of the internal angle bisectors. Prove that the area of the triangle $\triangle PQR$ cannot exceed $\frac{1}{4}$ of the area of the triangle $\triangle ABC$.*

Problem 113. *Let A, B, C be the angles of an acute triangle. Prove that:*

$$\sum_{cyc} \frac{\cos^3 A}{\sin B \sin C} \geq \frac{1}{2}$$

Problem 114. *Let $\triangle ABC$ be a triangle with side lengths a, b, c and circumradius R. Prove that:*

$$a^2 + b^2 + c^2 \leq 7R^2 + \frac{2(a+b+c)^2}{27}$$

Problem 115. *Let $\triangle ABC$ be a triangle with side lengths a, b, c and circumradius R such that $a + b + c = 3$. Find the minimum value of the following expression:*

$$\frac{1}{a^2} + \frac{1}{b^2} + \frac{1}{c^2} + \frac{1}{3R^2}$$

Problem 116. *Let $\triangle ABC$ be a triangle with side lengths a, b, c and circumradius R and inradius r such that $a + b + c = 3$. Find the minimal value of the following expression:*

$$\frac{1}{R^2} + \frac{1}{r^2} - \frac{6R}{r}$$

Problem 117. *Let $\triangle ABC$ be a triangle with side lengths a, b, c and m_a, m_b and m_c the lengths of the medians. Prove that:*

$$\frac{3}{4} \leq \frac{m_a + m_b + m_c}{a + b + c} \leq 1$$

Problem 118. *Let $\triangle ABC$ be a triangle with side lengths a, b, c and m_a, m_b and m_c the lengths of the medians, circumradius R and inradius r. Prove that:*

$$9r \leq m_a + m_b + m_c \leq 4R + r$$

Problem 119. *Let $\triangle ABC$ be a triangle with side lengths a, b, c and m_a, m_b and m_c the lengths of the medians. Prove that:*

$$\frac{m_a}{a} + \frac{m_b}{b} + \frac{m_c}{c} \geq \frac{3\sqrt{3}}{2}$$

Problem 120. *Let $\triangle ABC$ be a triangle with side lengths a, b, c and h_a, h_b and h_c the lengths of the altitudes. Prove that:*

$$\frac{h_a}{b+c} + \frac{h_b}{c+a} + \frac{h_c}{a+b} \leq \frac{3\sqrt{3}}{4}$$

Problem 121. *Let $\triangle ABC$ be a triangle with side lengths a, b, c and h_a, h_b and h_c the lengths of the altitudes, circumradius R and inradius r. Prove that:*

$$9r \leq h_a + h_b + h_c \leq 2R + 5r$$

Problem 122. *Let $\triangle ABC$ be a triangle with side lengths a, b, c and h_a, h_b and h_c the lengths of the altitudes. Prove that:*

$$h_a + h_b + h_c \leq \frac{\sqrt{3}(a+b+c)}{2}$$

Problem 123. *Prove that in any triangle, the length of the largest altitude is always smaller than twice the length of the smallest median.*

Problem 124. *Let $\triangle ABC$ be a triangle of inradius r and and $r_{medians}$ the inradius of the triangle with side lengths equal to the lengths of the medians. Prove that:*

$$\frac{3}{4} < \frac{r_{medians}}{r} < 1$$

Problem 125. *Let $\triangle_1$ and $\triangle_2$ be two triangles with sides (a, b, c) and $(\sqrt{a}, \sqrt{b}, \sqrt{c})$ respectively. Note by S_1 and S_2 their respective areas. Prove that:*

$$\frac{S_2}{\sqrt{S_1}} \geq \frac{\sqrt[4]{3}}{2}$$

Problem 126. *Let $\triangle ABC$ be a triangle with side lengths a, b and c. Let m_a and l_a be the lengths of the median and internal bisector issued from A. Prove that:*

$$0 \leq m_a - l_a \leq \frac{|b-c|}{2}$$

Problem 127. *Let $\triangle ABC$ be a triangle with circumcenter O and incenter I. Prove that the difference between the lengths of any two sides cannot exceed $2.OI$.*

Problem 128. *Let $\triangle ABC$ be a triangle with side lengths a, b and c. Let l_a, l_b, l_c be the lengths of the internal bisectors and m_a, m_b, m_c the lengths of the medians. Prove that:*

$$(m_a - l_a)^2 + (m_b - l_b)^2 + (m_c - l_c)^2 \leq \frac{1}{4}((a-b)^2 + (b-c)^2 + (c-a)^2)$$

Problem 129. *Let $\triangle ABC$ be a triangle with side lengths a, b and c. Let O be the circumcenter and I the incenter. Prove that:*

$$(a-b)^2 + (b-c)^2 + (c-a)^2 \leq 8.OI^2$$

Problem 130. *Let $\triangle ABC$ be a triangle with side lengths a, b and c. Let O be the circumcenter and I the incenter. Prove that:*

$$(a - b)^4 + (b - c)^4 + (c - a)^4 \leq 32.OI^4$$

Problem 131. *Let $\triangle ABC$ be a triangle with side lengths a, b and c. Let m_a, m_b, m_c be the lengths of the medians. Prove that:*

$$\frac{m_a}{a} + \frac{m_b}{b} + \frac{m_c}{c} \geq \frac{3(m_a + m_b + m_c)}{a + b + c}$$

Problem 132. *Let $\triangle ABC$ be a triangle with side lengths a, b and c, a circumradius R and inradius r. Prove that:*

$$\frac{a}{b + c} + \frac{b}{c + a} + \frac{c}{a + b} \leq \frac{R}{6r} + \frac{7}{6}$$

Problem 133. *Let $\triangle ABC$ be a triangle with m_a, m_b, m_c be the lengths of the medians and R the circumradius. Prove that:*

$$m_a^2 + m_b^2 + m_c^2 \leq \frac{27R^2}{4}$$

In particular, if the $\triangle ABC$ is an acute triangle, we have:

$$6R^2 \leq m_a^2 + m_b^2 + m_c^2 \leq \frac{27R^2}{4}$$

Problem 134. *Let $\triangle ABC$ be an acute triangle with side lengths a, b, c, circumcenter O and incenter I. Prove that:*

$$OI < \frac{a + b + c}{4}$$

Problem 135. *Let $\triangle ABC$ be an acute triangle with side lengths a, b, c, circumcenter O and incenter I. Prove that:*

$$4.OI^2 \leq (a - b)^2 + (b - c)^2 + (c - a)^2 \leq 8.OI^2$$

Problem 136. *Let $\triangle ABC$ be an acute triangle with side lengths a, b and c. Let h_a, h_b, h_c be the lengths of the altitudes. Prove that:*

$$\frac{1}{2} < \frac{h_a^2 + h_b^2 + h_c^2}{a^2 + b^2 + c^2} \leq \frac{3}{4}$$

Problem 137. *Let $\triangle ABC$ be a triangle with side lengths a, b and c. Let l_a, l_b, l_c be the lengths of the internal bisectors. Prove that:*

$$\frac{1}{2} < \frac{l_a^2 + l_b^2 + l_c^2}{a^2 + b^2 + c^2} \leq \frac{3}{4}$$

Problem 138. *Let $\triangle ABC$ be an acute triangle. Let s be the semiperimter, R the circumradius and r the inradius. Prove that:*

$$2R + r \leq s$$

Problem 139. *Let* $\triangle ABC$ *be an acute triangle with side lengths* a, b *and* c *and let* R *be the circumradius. Prove that:*

$$8R^2 \le a^2 + b^2 + c^2 \le 9R^2$$

Problem 140. *Let* $\triangle ABC$ *be an acute triangle with side lengths* a, b *and* c *and let* R *be the circumradius and* r *the inradius. Prove that:*

$$\begin{cases} ab + bc + ca \ge 4R^2 + 8Rr + 2r^2 \\ abc \ge 4Rr(r + 2R) \end{cases}$$

Problem 141. *Let* $\triangle ABC$ *be a triangle with side lengths* a, b *and* c *and let* R *the circumradius and* r *the inradius. Prove that:*

$$\frac{a^2}{b^2 + c^2} + \frac{b^2}{c^2 + a^2} + \frac{c^2}{a^2 + b^2} \le \frac{3R}{4r}$$

Problem 142. *Let* $\triangle ABC$ *be a triangle with medians of lengths* m_a, m_b, m_c. *Prove that:*

$$am_a + bm_b + cm_c \le \frac{3(a^3 + b^3 + c^3 + 3abc)}{4\sqrt{a^2 + b^2 + c^2}}$$

Problem 143. *Let* $\triangle ABC$ *be an acute triangle with side lengths* a, b *and* c *and let* R *be the circumradius and* r *the inradius. Prove that:*

$$ab(a + b) + bc(b + c) + ca(c + a) \ge 16R^3 + 16R^2r + 12Rr^2 + 4r^3$$

Problem 144. *Show that in any acute triangle, the circumradius cannot exceed* $\frac{2}{3}$ *the largest median.*

Problem 145. *Let* $\triangle ABC$ *be an acute triangle. Let* R *be the circumradius and* r *the inradius and* h_a, h_b, h_c *the lengths of the altitudes. Prove that:*

$$h_a h_b + h_b h_c + h_c h_a \ge 8r(R + r)$$

Problem 146. *Let* $\triangle ABC$ *be a triangle with side lengths* a, b *and* c. *Let* R, P, Q *be three points on the sides* AB, BC *and* AC *respectively. Prove that:*

$$\frac{PQ}{a + b} + \frac{QR}{b + c} + \frac{RP}{c + a} < 2$$

Problem 147. *Let* $\triangle ABC$ *be a triangle with side lengths* a, b *and* c. *If we note by* r *the inradius, then prove that:*

$$\frac{1}{a^2 + b^2} + \frac{1}{b^2 + c^2} + \frac{1}{c^2 + a^2} \ge \frac{81\sqrt{3}r}{(a + b + c)^3}$$

Problem 148. *Let* $\triangle ABC$ *be a triangle with side lengths* a, b, c *and medians of lengths* m_a, m_b, m_c. *If we note by* r *the inradius, prove that:*

$$\frac{m_a}{b + c} + \frac{m_b}{c + a} + \frac{m_c}{a + b} \le \frac{a + b + c}{8r}$$

Problem 149. *Let $\triangle ABC$ be a triangle with side lengths a, b and c. If we note by m_a, m_b and m_c the lengths of the medians, then prove that:*

$$\frac{m_a m_b}{b^2 + c^2} + \frac{m_b m_c}{c^2 + a^2} + \frac{m_c m_a}{a^2 + b^2} \le \frac{9}{8}$$

Problem 150. *Let $\triangle ABC$ be a triangle with side lengths a, b and c. If we note by r the inradius, then prove that:*

$$\frac{a^2 + b^2}{a + b} + \frac{b^2 + c^2}{b + c} + \frac{c^2 + a^2}{c + a} \le \frac{(a + b + c)^2}{6\sqrt{3}r}$$

4.2   Challenges

Challenge 48. *Let $\triangle ABC$ be an acute triangle with side lengths a, b and c. If we note by R and r the circumradius and inradius respectively, then prove:*

$$\frac{11R + 86r}{4r} \le \frac{(a + b + c)^3}{abc} \le \frac{8R}{r} + 11$$

Challenge 49. *Let $\triangle ABC$ a triangle with side lengths a, b and c. If we note by R and r the circumradius and inradius respectively, then prove:*

$$\frac{5(a^2 + b^2 + c^2)}{ab + bc + ca} \le \frac{R}{r} + 3$$

Challenge 50. *Let $\triangle ABC$ be a triangle with side lengths a, b, c, circumradius R and inradius r such that $a + b + c = 3$. Find the minimum value of the following expression:*

$$\frac{1}{R^2} + \frac{1}{r^2} - 4\left(\frac{1}{a^2} + \frac{1}{b^2} + \frac{1}{c^2}\right)$$

Challenge 51. *Let $\triangle ABC$ be a triangle with side lengths a, b, c, circumradius R and inradius r. Let m_a, m_b and m_c be the medians from A, B and C respectively. Prove that:*

$$0 \le \frac{m_a}{a} + \frac{m_b}{b} + \frac{m_c}{c} - \frac{3\sqrt{3}}{2} \le \frac{R}{r} - 2$$

Challenge 52. *Let $\triangle ABC$ be a triangle with side lengths a, b, c, semiperimiter s, circumradius R and inradius r. Let m_a, m_b and m_c be the medians from A, B and C respectively. Prove that:*

$$0 \le \frac{s}{2r} - \left(\frac{m_a}{a} + \frac{m_b}{b} + \frac{m_c}{c}\right) \le \frac{1}{2}\left(\frac{R}{r} - 2\right)$$

Challenge 53. *Let $\triangle ABC$ be a triangle of perimeter p. Let $R_{medians}$ and $r_{medians}$ be the circumradius and the inradius of the triangle formed by the medians. Prove that:*

$$2r_{medians} \le \frac{p}{6} \le R_{medians}$$

Challenge 54. *Let Δ_1 and Δ_2 be two triangles with sides (a, b, c) and $(a+b, b+c, a+c)$ respectively. Note by R_1 and R_2 their circumradii and r_1 and r_2 the inradii. Prove that:*

$$\frac{R_2}{r_2} \le \frac{R_1}{2r_1} + 1$$

Challenge 55. *Let* $\triangle ABC$ *be a triangle with side lengths* a, b, c. *Let* l_a, l_b, l_c *be the lengths of the internal bisectors of the angles of the triangle. Prove that:*

$$\frac{1}{2} < \frac{l_a + l_b + l_c}{a + b + c} \leq \frac{\sqrt{3}}{2}$$

Challenge 56. *Let* $\triangle ABC$ *be a triangle with side lengths* a, b, c, *semiperimiter* s, *circumradius* R *and inradius* r. *Let* l_a, l_b, l_c *be the lengths of the internal bisectors of the angles of the triangle. Prove that:*

$$0 \leq \frac{s}{2r} - \left(\frac{l_a}{a} + \frac{l_b}{b} + \frac{l_c}{c} \right) \leq \frac{2}{3} \left(\frac{R}{r} - 2 \right)$$

Challenge 57. *Let* $\triangle ABC$ *be a triangle with side lengths* a, b *and* c. *Let* m_a, m_b, m_c *be the lengths of the medians and* l_a, l_b, l_c *the lengths of the internal bisectors of the angles of the triangle. Prove that:*

$$0 \leq \frac{m_a - l_a}{a} + \frac{m_b - l_b}{b} + \frac{m_c - l_c}{c} < 1$$

Challenge 58. *Let* p *be the probability that the lengths of the altitudes of a triangle form the side lengths of a triangle. Prove that:*

$$p > \frac{3}{4}$$

Challenge 59. *Let* $\triangle ABC$ *be a triangle such that the lengths of the altitudes form the lengths of the sides of a triangle* $\triangle_{altitudes}$. *If we note by* r *and* $r_{altitudes}$ *the inradius of the two triangles respectively, prove that:*

$$r_{altitudes} \leq \frac{\sqrt{3}r}{2}$$

Challenge 60. *Let* $\triangle ABC$ *be a triangle with circumradius* R *and inradius* r *such that the lengths of the altitudes form the lengths of the sides of a triangle* $\triangle_{altitudes}$. *If we note by* A *and* $A_{altitudes}$ *the areas of the two triangles respectively, prove that:*

$$0 \leq \frac{3}{4} - \frac{A_{altitudes}}{A} \leq \frac{3}{2} \left(\frac{R}{r} - 2 \right)$$

Challenge 61. *Let* $\triangle ABC$ *be a triangle with circumradius* R *and inradius* r *such that the lengths of the internal bisectors form the lengths of the sides of a triangle* $\triangle_{bisectors}$. *If we note by* A *and* $A_{bisectors}$ *the areas of the two triangles respectively, prove that:*

$$0 \leq \frac{3}{4} - \frac{A_{bisectors}}{A} \leq \frac{R}{r} - 2$$

Challenge 62. *Let* $\triangle ABC$ *be a triangle with* m_a, m_b, m_c *be the lengths of the medians and* h_a, h_b, h_c *be the lengths of the altitudes. If we note by* O *and* I *the circumcenter and the incenter respectively, prove that:*

$$(m_a - h_a)^2 + (m_b - h_b)^2 + (m_c - h_c)^2 \leq 2.OI^2$$

Challenge 63. *Let $\triangle ABC$ be a triangle with h_a, h_b, h_c the lengths of the altitudes. If we note by O and I the circumcenter and incenter respectively, prove that:*

$$(h_a - h_b)^2 + (h_b - h_c)^2 + (h_c - h_a)^2 \leq 8.OI^2$$

Challenge 64. *Let $\triangle ABC$ be a triangle with l_a, l_b, l_c the lengths of the internal bisectors of the angles of the triangle. If we note by O and I the circumcenter and incenter respectively, prove that:*

$$(l_a - l_b)^2 + (l_b - l_c)^2 + (l_c - l_a)^2 \leq 8 \cdot OI^2$$

Challenge 65. *Let $\triangle ABC$ be a triangle with m_a, m_b, m_c the lengths of the medians. If we note by O and I the circumcenter and the incenter respectively, prove that:*

$$(m_a - m_b)^2 + (m_b - m_c)^2 + (m_c - m_a)^2 \leq \frac{9}{2}.OI^2$$

Challenge 66. *Let $\triangle ABC$ be a triangle with circumcenter O and incenter I.*

(i) *Prove that the difference between the lengths of any two medians is less or equal to $\sqrt{3}.OI$.*

(ii) *Prove that the difference between the lengths of any two altitudes is less or equal to $2.OI$.*

(iii) *Prove that the difference between the lengths of any two internal bisectors is less or equal to $2.OI$.*

Challenge 67. *Let $\triangle ABC$ be a triangle with circumcenter O and incenter I. Prove that:*

(i) *The difference between the lengths of any median and any altitude cannot exceed $2.OI$.*

(ii) *The difference between the lengths of a median and an altitude issued from the same point cannot exceed OI.*

Challenge 68. *Let $\triangle ABC$ be a triangle with side lengths a, b and c. Let m_a, m_b, m_c be the lengths of the medians from A, B and C respectively. Prove that:*

$$(m_a - m_b)^2 + (m_b - m_c)^2 + (m_c - m_a)^2 \leq \frac{9}{4}\left((a - b)^2 + (b - c)^2 + (c - a)^2\right)$$

Challenge 69. *Let $\triangle ABC$ be a triangle with side lengths a, b and c. Let l_a, l_b, l_c be the lengths of the internal bisectors of the angles of the triangle. Prove that:*

$$(l_a - l_b)^2 + (l_b - l_c)^2 + (l_c - l_a)^2 \leq 2((a - b)^2 + (b - c)^2 + (c - a)^2)$$

Challenge 70. *Let $\triangle ABC$ be a triangle with side lengths a, b and c. Let h_a, h_b, h_c be the lengths of the altitudes. Prove that:*

$$(h_a - h_b)^2 + (h_b - h_c)^2 + (h_c - h_a)^2 \leq (a - b)^2 + (b - c)^2 + (c - a)^2$$

Challenge 71. *Let $\triangle ABC$ be a triangle with side lengths a, b and c. Let m_a, m_b, m_c be the lengths of the medians and h_a, h_b, h_c be the lengths of the altitudes. Prove that:*

$$(m_a - h_a)^2 + (m_b - h_b)^2 + (m_c - h_c)^2 \leq \frac{9}{4}\left((a - b)^2 + (b - c)^2 + (c - a)^2\right)$$

Challenge 72. *Let $\triangle ABC$ be a triangle with side lengths a, b and c. Let l_a, l_b, l_c be the lengths of the internal bisectors of the angles of the triangle and h_a, h_b, h_c be the lengths of the altitudes. Prove that:*

$$(l_a - h_a)^2 + (l_b - h_b)^2 + (l_c - h_c)^2 \leq 2((a-b)^2 + (b-c)^2 + (c-a)^2)$$

Challenge 73. *Let $\triangle ABC$ be a triangle with side lengths a, b and c. Let m_a, m_b, m_c be the lengths of the medians and l_a, l_b, l_c be the lengths of internal bisectors of the angles of the triangle. Prove that:*

$$0 \leq \frac{m_a - l_a}{a} + \frac{m_b - l_b}{b} + \frac{m_c - l_c}{c} \leq \frac{1}{3}\left(\frac{R}{r} - 2\right)$$

Challenge 74. *Let $\triangle ABC$ be a triangle with side lengths a, b and c. Let m_a, m_b, m_c be the lengths of the medians and l_a, l_b, l_c be the lengths of internal bisectors of the angles of the triangle. If we note by O and I the circumcenter and incenter respectively, prove that:*

$$\frac{m_a - l_a}{a} + \frac{m_b - l_b}{b} + \frac{m_c - l_c}{c} \leq \frac{4.OI}{a+b+c}$$

Challenge 75. *Let $\triangle ABC$ be a triangle. Let m_a, m_b, m_c be the lengths of the medians, h_a, h_b, h_c be the lengths of the altitudes, l_a, l_b, l_c be the lengths of the internal bisectors. Prove that:*

$$(m_a + m_b + m_c)(h_a + h_b + h_c) < 2(l_a + l_b + l_c)^2$$

Challenge 76. *Let $\triangle ABC$ be a triangle. Let m_a, m_b, m_c be the lengths of the medians, h_a, h_b, h_c be the lengths of the altitudes, l_a, l_b, l_c be the lengths of the internal bisectors. Prove that:*

$$1 < \frac{m_a + m_b + m_c + h_a + h_b + h_c}{l_a + l_b + l_c} < 3$$

Challenge 77. *Let $\triangle ABC$ be a triangle with side lengths a, b and c. Let m_a, m_b, m_c be the lengths of the medians. If we note by O and I the circumcenter and incenter respectively, then prove that:*

$$\frac{(b-c)^2}{m_a^2} + \frac{(c-a)^2}{m_b^2} + \frac{(a-b)^2}{m_c^2} \leq \frac{64.OI^2}{a^2 + b^2 + c^2}$$

Challenge 78. *Let $\triangle ABC$ be a triangle with side lengths a, b and c. Let m_a, m_b, m_c be the lengths of the medians. Prove that:*

$$5\left(1 - \frac{ab + bc + ca}{a^2 + b^2 + c^2}\right) \leq \frac{(b-c)^2}{m_a^2} + \frac{(c-a)^2}{m_b^2} + \frac{(a-b)^2}{m_c^2} \leq 30\left(1 - \frac{ab + bc + ca}{a^2 + b^2 + c^2}\right)$$

Challenge 79. *Let $\triangle ABC$ be a triangle with side lengths a, b and c. Let h_a, h_b, h_c be the lengths of the altitudes. Prove that:*

$$\frac{(b-c)^2}{h_a^2} + \frac{(c-a)^2}{h_b^2} + \frac{(a-b)^2}{h_c^2} \geq 7\left(1 - \frac{ab + bc + ca}{a^2 + b^2 + c^2}\right)$$

Challenge 80. *Let $\triangle ABC$ be a triangle with side lengths a, b and c. Let l_a, l_b, l_c be the lengths of the internal bisectors of the angles of the triangle. Prove that:*

$$\frac{(b-c)^2}{l_a^2} + \frac{(c-a)^2}{l_b^2} + \frac{(a-b)^2}{l_c^2} \geq 6\left(1 - \frac{ab + bc + ca}{a^2 + b^2 + c^2}\right)$$

Challenge 81. *Let* $\triangle ABC$ *be a triangle with side lengths* a, b *and* c. *Let* m_a, m_b, m_c *be the lengths of the medians. Prove that:*

$$\frac{m_a + m_b - m_c}{a + b - c} + \frac{m_b + m_c - m_a}{b + c - a} + \frac{m_c + m_a - m_b}{c + a - b} \geq \frac{3\sqrt{3}}{2}$$

Challenge 82. *Let* $\triangle ABC$ *be a triangle with side lengths* a, b *and* c. *Let* l_a, l_b, l_c *be the lengths of the internal bisectors of the angles of the triangle. Prove that:*

$$\frac{l_a + l_b - l_c}{a + b - c} + \frac{l_b + l_c - l_a}{b + c - a} + \frac{l_c + l_a - l_b}{c + a - b} \geq \frac{3\sqrt{3}}{2}$$

Challenge 83. *Let* $\triangle ABC$ *be a triangle with side lengths* a, b *and* c. *Let* h_a, h_b, h_c *be the lengths of the altitudes. Prove that:*

$$\frac{h_a + h_b - h_c}{a + b - c} + \frac{h_b + h_c - h_a}{b + c - a} + \frac{h_c + h_a - h_b}{c + a - b} \geq \frac{3\sqrt{3}}{2}$$

Challenge 84. *Let* $\triangle ABC$ *be a triangle with side lengths* a, b *and* c. *Let* m_a, m_b, m_c *be the lengths of the medians from* A, B *and* C *respectively. If we note by* s *the semiperimiter and* r *the inradius, prove that:*

$$\frac{m_a + m_b - m_c}{a + b - c} + \frac{m_b + m_c - m_a}{b + c - a} + \frac{m_c + m_a - m_b}{c + a - b} \geq \frac{s}{2r}$$

Challenge 85. *Let* $\triangle ABC$ *be a triangle with side lengths* a, b *and* c. *Let* l_a, l_b, l_c *be the lengths of the internal bisectors of the angles of the triangle. If we note by* s *the semiperimiter and* r *the inradius, prove that:*

$$\frac{l_a + l_b - l_c}{a + b - c} + \frac{l_b + l_c - l_a}{b + c - a} + \frac{l_c + l_a - l_b}{c + a - b} \geq \frac{s}{2r}$$

Challenge 86. *Let* $\triangle ABC$ *be a triangle with side lengths* a, b *and* c. *Let* h_a, h_b, h_c *be the lengths of the altitudes. If we note by* s *the semiperimiter and* r *the inradius, prove that:*

$$\frac{h_a + h_b - h_c}{a + b - c} + \frac{h_b + h_c - h_a}{b + c - a} + \frac{h_c + h_a - h_b}{c + a - b} \geq \frac{s}{2r}$$

Challenge 87. *Let* $\triangle ABC$ *be a triangle with circumcenter* O *and incenter* I. *If we note by* m_a, l_a *and* h_a *the lengths of the median, internal bisector and altitude issued from* A, *prove that:*

$$m_a + l_a \leq 2h_a + \frac{5.OI}{4}$$

Challenge 88. *Let* $\triangle ABC$ *be a triangle with side lengths* a, b *and* c, *a circumcenter* O *and incenter* I. *Prove that:*

$$\frac{2a - b - c}{b + c} + \frac{2b - a - c}{c + a} + \frac{2c - a - b}{a + b} \leq \frac{4.OI}{a + b + c}$$

Challenge 89. *Let* $\triangle ABC$ *be a triangle with side lengths* a, b *and* c, *a circumcenter* O *and incenter* I. *Prove that:*

$$(ab + bc + ca)\left(\frac{1}{(a+b)^2} + \frac{1}{(b+c)^2} + \frac{1}{(c+a)^2}\right) \leq \frac{9}{4} + \frac{5.OI^2}{4(a^2 + b^2 + c^2)}$$

Challenge 90. *Consider an acute triangle $\triangle ABC$ with side lengths a, b and c. Prove that:*

$$A(a^2, b^2, c^2) \le \frac{4\sqrt{3}}{3} A^2(a, b, c)$$

where $A(x, y, z)$ is the area of a triangle with side lengths x, y and z.

Challenge 91. *Consider an acute triangle $\triangle ABC$ with side lengths a, b and c. Prove that:*

$$r(a^2, b^2, c^2) \le 2\sqrt{3} r^2(a, b, c)$$

and:

$$R(a^2, b^2, c^2) \ge \sqrt{3} R^2(a, b, c)$$

where $r(x, y, z)$ and $R(x, y, z)$ are respectively the inradius and circumradius of a triangle with side lengths x, y and z.

Challenge 92. *Let $\triangle ABC$ be an acute triangle with m_a, m_b, m_c be the lengths of the medians. If we note by O and I the circumcenter and the incenter respectively, prove that:*

$$2.OI^2 \le (m_a - m_b)^2 + (m_b - m_c)^2 + (m_c - m_a)^2 \le \frac{9}{2}.OI^2$$

Challenge 93. *Let $\triangle ABC$ be an acute non-equilateral triangle with side lengths a, b and c. Let m_a, m_b, m_c be the lengths of the medians. Prove that:*

$$\frac{1}{4} \le \frac{(m_a - m_b)^2 + (m_b - m_c)^2 + (m_c - m_a)^2}{(a - b)^2 + (b - c)^2 + (c - a)^2} \le 1$$

Challenge 94. *Let $\triangle ABC$ be an acute non-equilateral triangle with side lengths a, b and c. Let h_a, h_b, h_c be the lengths of the altitudes. Prove that:*

$$\frac{1}{2} \le \frac{(h_a - h_b)^2 + (h_b - h_c)^2 + (h_c - h_a)^2}{(a - b)^2 + (b - c)^2 + (c - a)^2} \le 1$$

Challenge 95. *Let $\triangle ABC$ be an acute non-equilateral triangle with side lengths a, b and c. Let l_a, l_b, l_c be the lengths of the internal bisectors. Prove that:*

$$\frac{2}{3} \le \frac{(l_a - l_b)^2 + (l_b - l_c)^2 + (l_c - l_a)^2}{(a - b)^2 + (b - c)^2 + (c - a)^2} \le 1$$

Challenge 96. *Let $\triangle ABC$ be a triangle. Let l_a, l_b, l_c be the lengths of the internal bisectors and m_a, m_b, m_c be the lengths of the medians. If we note by O and I the circumcenter and incenter respectively, prove that:*

$$0 \le (m_a^2 + m_b^2 + m_c^2) - (l_a^2 + l_b^2 + l_c^2) \le 3.OI^2$$

In particular, for an acute triangle:

$$\frac{5.OI^2}{3} \le (m_a^2 + m_b^2 + m_c^2) - (l_a^2 + l_b^2 + l_c^2) \le 3.OI^2$$

Challenge 97. *Let $\triangle ABC$ be a triangle with side lengths a, b and c. Let h_a, h_b, h_c be the lengths of the altitudes and m_a, m_b, m_c be the lengths of medians. If we note by O and I the circumcenter and incenter respectively, prove that:*

$$0 \le (m_a^2 + m_b^2 + m_c^2) - (h_a^2 + h_b^2 + h_c^2) \le \frac{13}{2}.OI^2$$

In particular, for an acute triangle:

$$2.OI^2 \le (m_a^2 + m_b^2 + m_c^2) - (h_a^2 + h_b^2 + h_c^2) \le \frac{13}{2}.OI^2$$

Challenge 98. *Show that in any acute triangle, the circumradius cannot exceed $\frac{2}{3}$ the lengths of the largest internal bisector.*

Challenge 99. *Let $\triangle ABC$ be a triangle, and R, P, Q three points on the sides AB, BC and AC respectively. If we note by r the inradius, prove that:*

$$PQ + QR + RP > 4r$$

Challenge 100. *Let $\triangle ABC$ be a triangle, and R, P, Q three points on the sides AB, BC and AC respectively. If we note by r, r_1, r_2, r_3 and r_4 the inradii of the triangles $\triangle ABC$, $\triangle AQR$, $\triangle BPR$, $\triangle CPQ$, and $\triangle RPQ$, prove that:*

$$r \le r_1 + r_2 + r_3 + r_4 < 3r$$

Challenge 101. *Let $\triangle ABC$ be a triangle, and R, P, Q three points on the sides AB, BC and AC respectively. Prove that:*

$$r_{PQR} \le r_{ABC}$$

where r_{XYZ} is the inradius of the triangle $\triangle XYZ$.

Challenge 102. *Let $\triangle ABC$ be a triangle, and R, P, Q three points on the sides AB, BC and AC respectively. Prove that:*

$$1 < \frac{PQ}{PC + CQ} + \frac{QR}{QA + AR} + \frac{RP}{RB + BP} < 3$$

Challenge 103. *Let $\triangle ABC$ be an acute triangle, and R, P, Q three points on the sides AB, BC and AC respectively. Prove that:*

$$\sqrt{2} < \frac{PQ}{PC + CQ} + \frac{QR}{QA + AR} + \frac{RP}{RB + BP} < 3$$

Challenge 104. *Let $\triangle ABC$ be an acute triangle with side lengths a, b and c, and R, P, Q three points on the sides AB, BC and AC respectively. Prove that:*

$$\frac{c - PQ}{a + b} + \frac{a - QR}{b + c} + \frac{b - RP}{c + a} \ge 0$$

Challenge 105. *Let $\triangle ABC$ be a triangle and R, P, Q three points on the sides AB, BC and AC respectively. If we note by m_a, m_b, m_c the lengths of the medians of $\triangle ABC$ and by m_p, m_q, m_r the lengths of the medians of $\triangle PQR$, prove that:*

$$m_p + m_q + m_r < \frac{4}{3}(m_a + m_b + m_c)$$

In particular, prove that we can find a triangle $\triangle ABC$ such that:

$$m_p + m_q + m_r > m_a + m_b + m_c$$

Challenge 106. *Let* $\triangle ABC$ *be a triangle and* R, P, Q *the feet of the internal angle bisectors. If we note by* r_{ABC} *and* r_{PQR} *the inradii of the triangles* $\triangle ABC$ *and* $\triangle PQR$ *respectively, prove that:*

$$\frac{1}{2} \leq \frac{r_{PQR}}{r_{ABC}} < \frac{2}{3}$$

Challenge 107. *Let* $\triangle ABC$ *be a triangle and* R, P, Q *the feet of the internal angle bisectors. If we note by* R_{ABC} *and* R_{PQR} *the circumradii of the triangles* $\triangle ABC$ *and* $\triangle PQR$ *respectively, prove that:*

$$R_{ABC} \geq 2R_{PQR}$$

Challenge 108. *Let* $\triangle ABC$ *be a triangle with circumcenter* O *and incenter* I. *Let* R, P, Q *be the feet of the internal angle bisectors. If we note by* p_{ABC} *and* p_{PQR} *the perimeters of the triangles* $\triangle ABC$ *and* $\triangle PQR$ *respectively, prove that:*

$$0 \leq \frac{p_{ABC}}{2} - p_{PQR} \leq 2.OI$$

Challenge 109. *Let* $\triangle ABC$ *be a triangle with circumcenter* O *and incenter* I. *Let* R, P, Q *be the feet of the internal angle bisectors. If we note by* R_{ABC} *and* R_{PQR} *the circumradii of* $\triangle ABC$ *and* $\triangle PQR$ *respectively, prove that:*

$$0 \leq R_{ABC} - 2R_{PQR} \leq OI$$

Challenge 110. *Let* $\triangle ABC$ *be a triangle with circumcenter* O *and incenter* I. *Let* R, P, Q *be the feet of the internal angle bisectors. If we note by* m_a, m_b *and* m_c *the lengths of the medians of the triangle* $\triangle ABC$ *and by* m_p, m_q *and* m_r *the lengths of the medians of the triangle* $\triangle PQR$, *prove that:*

$$0 \leq \frac{m_a + m_b + m_c}{2} - (m_p + m_q + m_r) \leq 2.OI$$

Challenge 111. *Let* $\triangle ABC$ *be a triangle with circumcenter* O *and incenter* I. *Let* R, P, Q *be the feet of the internal angle bisectors. If we note by* l_a, l_b *and* l_c *the internal bisectors of the triangle* $\triangle ABC$ *and by* l_p, l_q *and* l_r *the internal bisectors of the triangle* $\triangle PQR$, *prove that:*

$$0 \leq \frac{l_a + l_b + l_c}{2} - (l_p + l_q + l_r) \leq OI$$

Challenge 112. *Let* $\triangle ABC$ *be a triangle with circumcenter* O *and incenter* I. *Let* R, P, Q *be the feet of the internal angle bisectors. Prove that:*

$$(PQ - QR)^2 + (QR - RP)^2 + (RP - PQ)^2 \leq \frac{OI^2}{6}$$

Challenge 113. *Let* $\triangle ABC$ *be a triangle, and* R, P, Q *the feet of the internal angle bisectors. If we note by* r, r_1, r_2, r_3 *and* r_4 *the inradii of the triangles* $\triangle ABC$, $\triangle AQR$, $\triangle BPR$, $\triangle CPQ$, *and* $\triangle RPQ$, *prove that:*

$$2r \leq r_1 + r_2 + r_3 + r_4 < \frac{5r}{2}$$

Challenge 114. *Let* $\triangle ABC$ *be a triangle, and* R, P, Q *the feet of the internal angle bisectors respectively. If we note by* r_{ABC} *and* r_{AQR} *the inradii of the triangles* $\triangle ABC$, *and* $\triangle AQR$, *prove that:*

$$r_{ABC} \leq 3r_{AQR}$$

Challenge 115. *Let $\triangle ABC$ be a triangle with circumradius R and inradius r. Let I, J, K be the feet of the internal angle bisectors. Prove that:*

$$3\sqrt{3}r \leq IJ + JK + KI \leq \frac{3\sqrt{3}R}{2}$$

$$9r^2 \leq IJ.JK + JK.KI + KI.IJ \leq \frac{9Rr}{2}$$

$$3\sqrt{3}r^3 \leq IJ.JK.KI \leq \frac{3\sqrt{3}R^2r}{4}$$

Challenge 116. *Let $\triangle ABC$ be a triangle with circumradius R. Let D, E, F be the feet of the internal angle bisectors. If we note by O_1 and I_1 the circumcenter and incenter of the triangle $\triangle DEF$ respectively, prove that:*

$$O_1I_1 < \frac{R}{6}$$

Challenge 117. *Let $\triangle ABC$ be a triangle with circumradius R. Let D, E, F be the feet of the internal angle bisectors. Prove that there exists a real number $c < 1$ such that:*

$$\max(DE, EF, FD) < cR$$

Challenge 118. *Let $\triangle ABC$ be a triangle with inradius r. Let D, E, F be the feet of the internal angle bisectors. Prove that:*

$$\min(DE, EF, FD) > \frac{5r}{3}$$

Challenge 119. *Let $\triangle ABC$ be an acute triangle with side lengths a, b and c. Let R, P, Q be the feet of the altitudes on AB, BC and AC respectively. Prove that:*

$$\frac{PQ}{a+b} + \frac{QR}{b+c} + \frac{RP}{c+a} \leq \frac{3}{4}$$

Challenge 120. *Let $\triangle ABC$ be an acute triangle with circumcenter O and incenter I. Let R, P, Q be the feet of the altitudes on AB, BC and AC respectively. Prove that:*

$$(PQ - QR)^2 + (QR - RP)^2 + (RP - PQ)^2 \leq 12.OI^2$$

Challenge 121. *Let $\triangle ABC$ be an acute triangle with side lengths a, b and c. Let R, P, Q be the feet of the altitudes on AB, BC and AC respectively. Prove that:*

$$1 \leq \frac{PQ}{a+b-c} + \frac{QR}{b+c-a} + \frac{RP}{c+a-b} \leq \frac{3}{2}$$

Challenge 122. *Let $\triangle ABC$ be an acute triangle with side lengths a, b and c. Let R, P, Q be the feet of the altitudes on AB, BC and AC respectively. Prove that:*

$$1 < \frac{PQ}{c} + \frac{QR}{a} + \frac{RP}{b} \leq \frac{3}{2}$$

Challenge 123. *Let $\triangle ABC$ be an acute triangle with circumcenter O and incenter I. Let M_1, M_2, M_3 be the midpoints of the sides AB, BC and AC respectively, and H_1, H_2, H_3 the intersections of the altitudes with AB, BC and AC respectively. Prove that:*

$$2.OI^2 \leq M_1H_1^2 + M_2H_2^2 + M_3H_3^2 \leq \frac{25}{4}.OI^2$$

Challenge 124. *Let $\triangle ABC$ be an acute triangle with circumcenter O and incenter I. Let M_1, M_2, M_3 be the midpoints of the sides AB, BC and AC respectively, and H_1, H_2, H_3 the intersections of the altitudes with AB, BC and AC respectively. Prove that:*

$$2.OI \leq M_1H_1 + M_2H_2 + M_3H_3 \leq 4.OI$$

Challenge 125. *Let $\triangle ABC$ be an acute triangle with side lengths a, b and c, a circumcenter O and an incenter I. Let M_1, M_2, M_3 be the midpoints of the sides AB, BC and AC respectively, and H_1, H_2, H_3 the intersections of the altitudes with AB, BC and AC respectively. Prove that:*

$$\frac{4.OI}{a+b+c} \leq \frac{M_1H_1}{c} + \frac{M_2H_2}{a} + \frac{M_3H_3}{b} \leq \frac{13.OI}{a+b+c}$$

Challenge 126. *Let m_a, m_b, m_c be the lengths of the medians of an acute triangle and R and r are respectively the circumradius and inradius. Prove that:*

$$4(r^2 + rR + R^2) < am_a + bm_b + cm_c \leq \frac{9\sqrt{3}R^2}{2}$$

Challenge 127. *Let $\triangle ABC$ be a triangle with side lengths a, b and c, a circumcenter O and incenter I. If we note by α, β, γ the angles opposite the sides, then prove:*

$$0 \leq \cos^2 \alpha + \cos^2 \beta + \cos^2 \gamma - \frac{3}{4} \leq \frac{34.OI^2}{(a+b+c)^2}$$

Challenge 128. *Let $\triangle ABC$ be a triangle with side lengths a, b and c, a circumcenter O and incenter I. If we note by α, β, γ the angles opposite the sides, then prove:*

$$0 \leq \frac{a+b+c}{2} - (a\cos\alpha + b\cos\beta + c\cos\gamma) \leq 2.OI$$

Challenge 129. *Let $\triangle ABC$ be a triangle with side lengths a, b and c, a circumcenter O and incenter I. If we note by α, β, γ the angles opposite the sides, then prove:*

$$0 \leq \frac{1}{8} - \cos\alpha.\cos\beta.\cos\gamma \leq \frac{17.OI^2}{(a+b+c)^2}$$

Challenge 130. *Let $\triangle ABC$ be a triangle with side lengths a, b and c, a circumcenter O and incenter I. If we note by α, β, γ the angles opposite the sides, then prove:*

$$0 \leq \frac{9}{4(a+b+c)} - \left(\frac{\cos\alpha}{b+c} + \frac{\cos\beta}{c+a} + \frac{\cos\gamma}{a+b}\right) \leq \frac{5.OI}{(a+b+c)^2}$$

Challenge 131. *Let $\triangle ABC$ be a triangle with side lengths a, b and c, a circumcenter O and incenter I. If we note by α, β, γ the angles opposite the sides, then prove:*

$$(b-c)^2\cos\alpha + (c-a)^2\cos\beta + (a-b)^2\cos\gamma \leq \frac{OI(a+b+c)}{2}$$

Challenge 132. *Let $\triangle ABC$ be a triangle with side lengths a, b and c, a circumcenter O and incenter I. If we note by α, β, γ the angles opposite the sides, then prove:*

$$0 \leq \sum_{cyc}(b+c-a).\left(\cos\alpha - \frac{1}{2}\right) \leq 2.OI$$

Challenge 133. *Let $\triangle ABC$ be a triangle with side lengths a, b and c and α, β, γ the angles opposite the sides. Prove that:*

$$\cos\alpha + \cos\beta + \cos\gamma + \frac{5(a^2 + b^2 + c^2)}{2(ab + bc + ca)} \geq 4$$

Challenge 134. *Let $\triangle ABC$ be a triangle with side lengths a, b and c, and medians of lengths m_a, m_b and m_c. Let α, β, γ be the angles opposite the sides, and R and r the cicrcumradius and inradius respectively. Prove that:*

$$\frac{9r}{2} \leq m_a \cdot \cos\alpha + m_b \cdot \cos\beta + m_c \cdot \cos\gamma \leq \frac{9R}{4}$$

Challenge 135. *Let $\triangle ABC$ be a triangle with side lengths a, b and c, and medians of lengths m_a, m_b and m_c. Let α, β, γ be the angles opposite the sides, and R and r the circumradius and inradius respectively. Prove that:*

$$|(m_b - m_c) \cdot \cos\alpha + (m_c - m_a) \cdot \cos\beta + (m_a - m_b) \cdot \cos\gamma| \leq R - 2r$$

Challenge 136. *Let $\triangle ABC$ be a triangle and P, Q, R three points on the sides BC, AC and AB respectively such that (AP), (BQ) and (CR) are concurrent. If we note by A_1 and A_2 the areas of triangles $\triangle ABC$ and $\triangle PQR$ respectively, prove that:*

$$A_2 \leq \frac{A_1}{4}$$

Challenge 137. *In a triangle $\triangle ABC$ with side lengths a, b, c and medians of lengths m_a, m_b and m_c, prove that:*

$$\sum_{cyc} a(m_b + m_c - m_a) \geq \frac{\sqrt{3}(a^2 + b^2 + c^2)}{2}$$

Challenge 138. *Let $\triangle ABC$ be a triangle and P, Q, R on the sides BC, AC and AB respectively such that (AP), (BQ) and (CR) are concurrent. Prove that:*

$$\frac{AP}{BP} + \frac{BQ}{CQ} + \frac{CR}{AR} \geq 3\sqrt{3}$$

Challenge 139. *Let $\triangle ABC$ be a triangle and P, Q, R on the sides BC, AC and AB respectively. Prove that:*

$$\frac{AP^2}{BP.CP} + \frac{BQ^2}{CQ.AQ} + \frac{CR^2}{AR.BR} \geq 9$$

Challenge 140. *Let $\triangle ABC$ be an acute triangle with side lengths a, b and c, circumcenter O and incenter I, prove that:*

$$\left(\frac{1}{\cos A} - 1\right) \cdot \left(\frac{1}{\cos B} - 1\right) \cdot \left(\frac{1}{\cos C} - 1\right) \geq 1 + \frac{16.OI^2}{(a + b + c)^2}$$

Challenge 141. *Let $\triangle ABC$ be an acute triangle with side lengths a, b and c, circumradius R and inradius r. Prove that:*

$$a\left(\frac{1}{\cos A} - 2\right) + b\left(\frac{1}{\cos B} - 2\right) + c\left(\frac{1}{\cos C} - 2\right) \geq 25(R - 2r)$$

Challenge 142. *Let $\triangle ABC$ be a triangle and P, Q, R on the sides BC, AC and AB respectively. Prove that:*

$$(AB + BC + CA)^2 \geq 3(AP.BP + BQ.CQ + CR.AR)$$

Challenge 143. *Let $\triangle ABC$ be a triangle with side lengths a, b and c, a circumcenter O and incenter I. If we note by α, β, γ the angles opposite the sides, then prove:*

$$0 \leq \frac{3\sqrt{3}}{2} - (\sin \alpha + \sin \beta + \sin \gamma) \leq \frac{4.OI}{a + b + c}$$

Challenge 144. *Let $\triangle ABC$ be a triangle with side lengths a, b and c, a circumcenter O and incenter I. If we note by α, β, γ the angles opposite the sides, then prove:*

$$(b - c)^2 \sin \alpha + (c - a)^2 \sin \beta + (a - b)^2 \sin \gamma \leq 8.OI^2$$

Challenge 145. *Let $\triangle ABC$ be a triangle with side lengths a, b and c, a circumradius R and inradius r. If we note by α, β, γ the angles opposite the sides, then prove:*

$$\frac{\sqrt{3}}{2} - \frac{R}{2r} + 1 \leq \sum_{cyc} \left(1 - \frac{2a}{a + b + c}\right) \sin \alpha \leq \frac{\sqrt{3}}{2}$$

Challenge 146. *Let $\triangle ABC$ be a triangle with medians of lengths m_a, m_b and m_c. If we note by α, β, γ the angles opposite the sides and R the circumradius, then prove:*

$$m_a \sin \alpha + m_b \sin \beta + m_c \sin \gamma \leq \frac{9\sqrt{3}R}{4}$$

Challenge 147. *Let $\triangle ABC$ be an acute triangle with R and r the circumradius and inradius respectively. If we note by $\alpha_H, \beta_H, \gamma_H$ the angles opposite the sides of the orthic triangle, prove that:*

$$\cos^2 \alpha_H + \cos^2 \beta_H + \cos^2 \gamma_H + \frac{17}{4} \leq \frac{5R}{2r}$$

Challenge 148. *Let $\triangle ABC$ be an acute triangle with R and r the circumradius and inradius respectively. If we note by $\alpha_H, \beta_H, \gamma_H$ the angles opposite the sides of the orthic triangle, prove that:*

$$\sqrt{3}(\sin \alpha_H + \sin \beta_H + \sin \gamma_H) + \frac{4R}{r} \geq \frac{25}{2}$$

Challenge 149. *Let $\triangle ABC$ be an acute triangle with R and r the circumradius and inradius respectively. If we note by α, β, γ the angles opposite the sides of the triangle $\triangle ABC$, and $\alpha_H, \beta_H, \gamma_H$ the angles opposite the sides of the orthic triangle, prove that:*

$$1 \leq \frac{\cos \alpha + \cos \beta + \cos \gamma}{\cos \alpha_H + \cos \beta_H + \cos \gamma_H} \leq \frac{R}{r} - 1$$

Challenge 150. *Let $\triangle ABC$ be a triangle with centroid G. If we note by r, r_1, r_2 and r_3 the inradii of the triangles $\triangle ABC$, $\triangle GAB$, $\triangle GBC$ and $\triangle GCA$ respectively, prove that:*

$$r_1 + r_2 + r_3 \geq 3\sqrt{3}(2 - \sqrt{3})r$$

Chapter 5

Solutions

5.1 3-variable inequalities

5.1.1 Symmetric inequalities

Solution 1.

We will give three solutions to this inequality.

- ***First Solution:***

*Applying **Cauchy-Schwarz inequality**, we find that*

$$\sum_{cyc} \frac{a^3}{(1+a)(1+b)} \geq \frac{(a^2+b^2+c^2)^2}{\sum_{cyc} a(1+a)(1+b)}$$

Therefore, it remains to prove that

$$\frac{(a^2+b^2+c^2)^2}{\sum_{cyc} a(1+a)(1+b)} \geq \frac{a^2+b^2+c^2}{4}$$

which is equivalent to

$$\frac{a^2+b^2+c^2}{\sum_{cyc} a(1+a)(1+b)} \geq \frac{1}{4}$$

Moreover, notice that

$$\sum_{cyc} a(1+a)(1+b) = \sum_{cyc} (a+a^2+ab+a^2 b)$$

*Using **Chebyshev's** and **Rearrangement inequalities**, we get*

$$\sum_{cyc} a^2 \geq \frac{(a+b+c)^2}{3} = a+b+c$$

$$\sum_{cyc} a^2 \geq \sum_{cyc} ab$$

It is sufficient to prove that

$$a^2b + b^2c + c^2a \leq a^2 + b^2 + c^2$$

which is true since

$$(a+b+c)(a^2+b^2+c^2) - 3(a^2b+b^2c+c^2a) = \sum_{cyc} a(a-b)^2 \geq 0$$

Therefore

$$\sum_{cyc} a(1+a)(1+b) = \sum_{cyc} \left(a + a^2 + ab + a^2b\right)$$

$$\leq \sum_{cyc} \left(a^2 + a^2 + a^2 + a^2\right)$$

$$= 4\left(\sum_{cyc} a^2\right)$$

The problem is completely solved. Equality holds for $a = b = c = 1$.

• **Second Solution:**

Using **AM-GM inequality**, we get

$$\frac{(2+a+b)^2}{4} \geq (1+a)(1+b)$$

Therefore, we are going to prove the following stronger inequality

$$\frac{a^3}{(a+b+2)^2} + \frac{b^3}{(b+c+2)^2} + \frac{c^3}{(c+a+2)^2} \geq \frac{a^2+b^2+c^2}{16}$$

Applying **Titu's lemma**, we obtain

$$\sum_{cyc} \frac{a^3}{(a+b+2)^2} = \sum_{cyc} \frac{a^4}{a(a+b+2)^2} \geq \frac{(a^2+b^2+c^2)^2}{\sum_{cyc} a(a+b+2)^2}$$

Thus, it is enough to prove

$$16(a^2+b^2+c^2) \geq \sum_{cyc} a(a+b+2)^2$$

After homogenisation, we get

$$48(a^2+b^2+c^2)(a+b+c) \geq \sum_{cyc} a(5a+5b+2c)^2$$

which can be reduced to

$$\sum_{cyc}(23a^3 - 6a^2b + 3a^2c - 20abc) \geq 0$$

This last inequality is true since it can be transformed into

$$6\sum_{cyc}(a^3 - a^2b) + 17\sum_{cyc}(a^3 - abc) + 3\sum_{cyc}(a^2c - abc) \geq 0$$

which is obviously true by the **Rearrangement inequality.**

The proof is finished. Equality holds for $a = b = c = 1$.

- *Third Solution:*

Applying **AM-GM** *and* **Cauchy-Schwarz inequalities***, we deduce*

$$(1+a)(1+b) \leq \frac{(a+b+2)^2}{4}$$
$$\leq \frac{(a^2+b^2+2)(1+1+2)}{4}$$
$$= a^2 + b^2 + 2$$

(Alternatively: $a^2 + b^2 + 2 - (1+a)(1+b) = \frac{(2a-b-1)^2}{4} + \frac{3(b-1)^2}{4} \geq 0$)

Therefore, it is enough to prove that

$$\frac{a^3}{a^2+b^2+2} + \frac{b^3}{b^2+c^2+2} + \frac{c^3}{c^2+a^2+2} \geq \frac{a^2+b^2+c^2}{4}$$

Applying **Titu's lemma***, we get*

$$\sum_{cyc}\frac{a^3}{a^2+b^2+2} = \sum_{cyc}\frac{a^4}{a(a^2+b^2+2)} \geq \frac{(a^2+b^2+c^2)^2}{\sum_{cyc}a(a^2+b^2+2)}$$

Consequently, we only need to prove

$$\frac{(a^2+b^2+c^2)^2}{\sum_{cyc}a(a^2+b^2+2)} \geq \frac{a^2+b^2+c^2}{4}$$

which is equivalent to prove

$$4(a^2+b^2+c^2) \geq \sum_{cyc}a(a^2+b^2+2)$$

On the other hand, we have

$$4\left(\sum_{cyc}a^2\right) - \sum_{cyc}a(a^2+b^2+2) = \sum_{cyc}a^2 + \sum_{cyc}a^2b - 6$$
$$= \sum_{cyc}(b+a^2b) - 2\left(\sum_{cyc}ab\right)$$
$$\geq 2\left(\sum_{cyc}\sqrt{b.a^2b}\right) - 2\left(\sum_{cyc}ab\right)$$

$$= 2\left(\sum_{cyc} ab\right) - 2\left(\sum_{cyc} ab\right)$$

$$= 0$$

We are done. Equality holds for $a = b = c = 1$.

■

Solution 2.

We will give two solutions to this inequality.

• *First Solution:*

Notice that

$$a^3 + b^3 + c^3 + 3abc = \sum_{cyc} a(a^2 + bc)$$

Using **Cauchy-Schwarz inequality**, we find that

$$(a^3 + b^3 + c^3 + 3abc)^2 = \left(\sum_{cyc} a(a^2 + bc)\right)^2$$

$$\leq \left(\sum_{cyc} a^2\right)\left(\sum_{cyc} (a^2 + bc)^2\right)$$

$$= \left(\sum_{cyc} a^2\right)\left(\sum_{cyc} (a^4 + 2a^2 bc + b^2 c^2)\right)$$

$$\leq \left(\sum_{cyc} a^2\right)\left(\sum_{cyc} (a^4 + 2a^2 b^2) + \sum_{cyc} a^2 b^2\right)$$

$$\leq \left(\sum_{cyc} a^2\right)\left((a^2 + b^2 + c^2)^2 + \frac{(a^2 + b^2 + c^2)^2}{3}\right)$$

$$= \frac{4(a^2 + b^2 + c^2)^3}{3}$$

This ends the proof. Equality holds for $a = b = c$.

Comment 1.

The intuition behind this solution is simple: **Lagrange Multipliers**. In fact:

Let $s = a^2 + b^2 + c^2 \geq 0$ and we are looking to maximise

$$a^3 + b^3 + c^3 + 3abc$$

Let

$$L(a, b, c, \lambda) = a^3 + b^3 + c^3 + 3abc + \lambda(a^2 + b^2 + c^2 - s)$$

Finding the critical points, we get

$$\begin{cases} \frac{\partial L}{\partial a} = 3a^2 + 3bc + 2\lambda a = 0 \\ \frac{\partial L}{\partial b} = 3b^2 + 3ca + 2\lambda b = 0 \\ \frac{\partial L}{\partial c} = 3c^2 + 3ab + 2\lambda c = 0 \end{cases}$$

Thus

$$\frac{a^2 + bc}{a} = \frac{b^2 + ca}{b} = \frac{c^2 + ab}{c} = -\frac{2\lambda}{3}$$

*Instead of solving this system, we can apply the **Cauchy-Schwarz inequality** to the vectors (a, b, c) and $(a^2 + bc, b^2 + ca, c^2 + ab)$.*

- ***Second Solution:***

Let $a + b + c = 3u$, $ab + bc + ca = 3v^2$ and $abc = w^3$. We have

$$\begin{cases} a^2 + b^2 + c^2 = 9u^2 - 6v^2 \\ a^3 + b^3 + c^3 = 27u^3 - 27uv^2 + 3w^3 \end{cases}$$

Therefore, we can rewrite the original inequality as

$$f(w^3) = 4(3u^2 - 2v^2)^3 - (9u^3 - 9uv^2 + 2w^3)^2 \geq 0$$

*f is a decreasing function of w^3, hence by **uvw method**, it attains a minimal value when w^3 attains a maximal value, which happens when two variables are equal. WLOG, let $c = b$, and we get*

$$(a - b)^2(a^4 + 2a^3b + 9a^2b^2 + 4ab^3 + 20b^4) \geq 0$$

which is obviously true. Equality holds for $a = b = c$.

■

Solution 3.

We have four solutions to this problem.

- ***First Solution:***

We proved in the previous problem that

$$a^2 + b^2 + c^2 \geq 3$$

*Using **Lagrange's identity**, we find that*

$$(a^2 + b^2)(a^2 + c^2) = (a^2 + bc)^2 + a^2(b - c)^2 \geq (a^2 + bc)^2$$

Thus

$$\sqrt{(a^2 + b^2)(a^2 + c^2)} \geq a^2 + bc$$

*Applying **Cauchy-Schwarz inequality**, we conclude that*

$$\sum_{cyc} \frac{a^2 + b^2}{a + b} \geq \frac{\left(\sum_{cyc} \sqrt{a^2 + b^2}\right)^2}{2(a + b + c)}$$

$$= \frac{2(a^2 + b^2 + c^2) + 2\left(\sum_{cyc} \sqrt{a^2 + b^2}\sqrt{a^2 + c^2}\right)}{2(a + b + c)}$$

$$\geq \frac{2(a^2 + b^2 + c^2) + 2\left(\sum_{cyc}(a^2 + bc)\right)}{2(a + b + c)}$$

$$= \frac{3(a^2 + b^2 + c^2) + (a + b + c)^2}{2(a + b + c)}$$

$$\geq \frac{2\sqrt{3(a^2 + b^2 + c^2)}(a + b + c)^2}{2(a + b + c)}$$

$$= \sqrt{3(a^2 + b^2 + c^2)}$$

$$\geq 3$$

We are done. Equality at $a = b = c = 1$.

Comment 2.

*Instead of **Lagrange's identity**, we could have used **Complex Numbers** to prove that inequality. In fact, we have*

$$(a^2 + b^2)(a^2 + c^2) = |a + ib|^2|a - ic|^2$$
$$= |a^2 + bc + a(b - c)i|^2$$
$$= (a^2 + bc)^2 + a^2(b - c)^2$$
$$\geq (a^2 + bc)^2$$

• *Second Solution:*

From the previous problem, we know that

$$a^2 + b^2 + c^2 \geq 3$$

Thus, it is sufficient to prove

$$\frac{a^2 + b^2}{a + b} + \frac{b^2 + c^2}{b + c} + \frac{c^2 + a^2}{c + a} \geq \sqrt{3(a^2 + b^2 + c^2)}$$

*Applying **Bernoulli's inequality**, we obtain*

$$\sqrt{1 + x} \leq 1 + \frac{x}{2}$$

Therefore

$$\sqrt{3(a^2 + b^2 + c^2)} = \sqrt{(a + b + c)^2 + (b - c)^2 + (c - a)^2 + (a - b)^2}$$

$$= (a+b+c)\sqrt{1 + \frac{(b-c)^2 + (c-a)^2 + (a-b)^2}{(a+b+c)^2}}$$

$$\le (a+b+c)\left(1 + \frac{(b-c)^2 + (c-a)^2 + (a-b)^2}{2(a+b+c)^2}\right)$$

$$= a+b+c + \frac{(b-c)^2 + (c-a)^2 + (a-b)^2}{2(a+b+c)}$$

$$= \sum_{cyc}\left(\frac{a+b}{2} + \frac{(a-b)^2}{2(a+b+c)}\right)$$

$$\le \sum_{cyc}\left(\frac{a+b}{2} + \frac{(a-b)^2}{2(a+b)}\right)$$

$$= \sum_{cyc}\frac{(a+b)^2 + (a-b)^2}{2(a+b)}$$

$$= \sum_{cyc}\frac{a^2+b^2}{a+b}$$

as desired. Equality holds for $a = b = c = 1$.

- **Third Solution:**

From the previous problem, we know that

$$a^2 + b^2 + c^2 \ge 3$$

Applying **AM-GM inequality**, we get

$$2\sqrt{3(a^2+b^2+c^2)} \le \frac{3(a^2+b^2+c^2)}{a+b+c} + a+b+c$$

Let

$$L = \sum_{cyc}\frac{a^2+b^2}{a+b} - \frac{1}{2}\left[\frac{3(a^2+b^2+c^2)}{a+b+c} + a+b+c\right]$$

It is sufficient to prove $L \ge 0$ which we can easily do as follows

$$L = \sum_{cyc}\frac{a^2+b^2}{a+b} - \frac{1}{2}\left[\frac{3(a^2+b^2+c^2)}{a+b+c} + a+b+c\right]$$

$$= \sum_{cyc}\left(\frac{a^2+b^2}{a+b} - \frac{a+b}{2}\right) - \frac{1}{2}\left(\frac{3(a^2+b^2+c^2)}{a+b+c} - (a+b+c)\right)$$

$$= \sum_{cyc}\frac{(a-b)^2}{2(a+b)} - \frac{\sum_{cyc}(a-b)^2}{2(a+b+c)}$$

$$= \sum_{cyc}\frac{1}{2}\left(\frac{1}{a+b} - \frac{1}{a+b+c}\right)(a-b)^2$$

$$= \sum_{cyc}\frac{c(a-b)^2}{2(a+b)(a+b+c)}$$

$$\geq 0$$

as desired. Equality holds for $a = b = c = 1$.

- **Fourth Solution:**

From the previous problem, we know that

$$a^2 + b^2 + c^2 \geq 3$$

Thus, it is sufficient to prove

$$\frac{a^2 + b^2}{a + b} + \frac{b^2 + c^2}{b + c} + \frac{c^2 + a^2}{c + a} \geq \sqrt{3(a^2 + b^2 + c^2)}$$

Subtracting $a + b + c$ from both sides, we get

$$\sum_{cyc} \left(\frac{a^2 + b^2}{a + b} - \frac{a + b}{2} \right) \geq \sqrt{3(a^2 + b^2 + c^2)} - (a + b + c)$$

that can be written as

$$\sum_{cyc} \frac{(a - b)^2}{2(a + b)} \geq \frac{\sum_{cyc}(a - b)^2}{\sqrt{3(a^2 + b^2 + c^2)} + (a + b + c)}$$

On the other hand, we know that

$$\sqrt{3(a^2 + b^2 + c^2)} + a + b + c \geq 2(a + b + c)$$

Therefore, it is enough to prove

$$\sum_{cyc} \frac{(a - b)^2}{a + b} \geq \frac{\sum_{cyc}(a - b)^2}{a + b + c}$$

or equivalently

$$\frac{1}{a + b + c} \sum_{cyc} \frac{c}{a + b}(a - b)^2 \geq 0$$

which is true. Equality holds for $a = b = c = 1$.

Comment 3.

This is probably the most intuitive technique in solving inequalities problems. It is based on the most elementary property of real numbers:

$$\forall x \in \mathbb{R}: \qquad x^2 \geq 0$$

The basic idea is to rewrite the original inequality as a sum of squares. In the following examples, we will show how this technique could be used to solve some difficult problems.

○ **First example:** *[AoPS]*

Let a, b, c be positive real numbers. Prove that:

$$\frac{a^2}{b} + \frac{b^2}{c} + \frac{c^2}{a} \geq a + b + c$$

○ **Proof:**

We have

$$\sum_{cyc} \left(\frac{a^2}{b} - a \right) = \sum_{cyc} \left(\frac{a^2}{b} - 2a + b \right)$$

$$= \sum_{cyc} \frac{a^2 - 2ab + b^2}{b}$$

$$= \sum_{cyc} \frac{(a-b)^2}{b}$$

$$\geq 0$$

as desired. Equality holds for $a = b = c$.

○ **Second example:** *[AoPS]*

Let a, b, c be positive real numbers. Prove that:

$$\frac{a^2}{b^2+c^2} + \frac{b^2}{c^2+a^2} + \frac{c^2}{a^2+b^2} \geq \frac{a}{b+c} + \frac{b}{c+a} + \frac{c}{a+b}$$

○ **Proof:**

In fact, we have

$$\sum_{cyc} \left(\frac{a^2}{b^2+c^2} - \frac{a}{b+c} \right) = \sum_{cyc} \frac{ab(a-b) + ac(a-c)}{(b+c)(b^2+c^2)}$$

$$= \sum_{cyc} \left[\frac{ab(a-b)}{(b+c)(b^2+c^2)} - \frac{ab(a-b)}{(a+c)(a^2+c^2)} \right]$$

$$= (a^2+b^2+c^2+ab+bc+ca) \sum_{cyc} \frac{ab(a-b)^2}{(b+c)(a+c)(b^2+c^2)(a^2+c^2)}$$

$$\geq 0$$

The inequality is proved.

○ **Third example:** *[IMO 2005]*

Suppose that x, y, z are positive real numbers such that $xyz \geq 1$. Prove that:

$$\frac{x^5 - x^2}{x^5 + y^2 + z^2} + \frac{y^5 - y^2}{y^5 + z^2 + x^2} + \frac{z^5 - z^2}{z^5 + y^2 + x^2} \geq 0$$

○ **Proof:**

First of all, we have

$$\frac{x^5 - x^2}{x^5 + y^2 + z^2} - \frac{x^5 - x^2.xyz}{x^5 + (y^2 + z^2)xyz} = \frac{x^4(xyz - 1)(x^2 + y^2 + z^2)}{(x^5 + y^2 + z^2)(x^4 + yz(y^2 + z^2))} \geq 0$$

Consequently, we can rewrite the original inequality in standard equal-degree form

$$\sum_{cyc} \frac{x^5 - x^2}{x^5 + y^2 + z^2} \geq \sum_{cyc} \frac{x^5 - x^2.xyz}{x^5 + (y^2 + z^2)xyz}$$

$$= \sum_{cyc} \frac{x^4 - x^2yz}{x^4 + yz(y^2 + z^2)}$$

On the other hand, we have

$$\frac{x^4 - x^2yz}{x^4 + yz(y^2 + z^2)} - \frac{2x^4 - x^2(y^2 + z^2)}{2x^4 + (y^2 + z^2)^2} = \frac{x^4(y - z)^2(x^2 + y^2 + z^2)}{(x^4 + yz(y^2 + z^2))(2x^4 + (y^2 + z^2)^2)} \geq 0$$

Therefore

$$\frac{x^4 - x^2yz}{x^4 + yz(y^2 + z^2)} \geq \frac{2x^4 - x^2(y^2 + z^2)}{2x^4 + (y^2 + z^2)^2}$$

Let $a = x^2$, $b = y^2$ and $c = z^2$. We need to prove

$$\sum_{cyc} \frac{2a^2 - a(b + c)}{2a^2 + (b + c)^2} \geq 0$$

or equivalently

$$\sum_{cyc} (a - b) \left(\frac{a}{2a^2 + (b + c)^2} - \frac{b}{2b^2 + (c + a)^2} \right) \geq 0$$

that can be transformed into

$$\sum_{cyc} (a - b)^2 \frac{c^2 + c(a + b) + a^2 - ab + b^2}{(2a^2 + (b + c)^2)(2b^2 + (c + a)^2)} \geq 0$$

which is obviously true. The equality holds for $a = b = c$.

○ **Fourth example:** *[AoPS]*

Let $\triangle ABC$ be a triangle with side lengths a, b and c. Prove that:

$$\frac{6(a^2 + b^2 + c^2)}{a + b + c} \geq \frac{(a + b)^2}{b + c} + \frac{(b + c)^2}{c + a} + \frac{(c + a)^2}{a + b}$$

○ **Proof:**

After substracting $2(a + b + c)$ from both sides, we need to prove

$$\frac{2((a - b)^2 + (b - c)^2 + (c - a)^2)}{a + b + c} \geq \frac{(c - a)^2}{b + c} + \frac{(a - b)^2}{c + a} + \frac{(b - c)^2}{a + b}$$

that can be rewritten as

$$\sum_{cyc} \left(\frac{2}{a+b+c} - \frac{1}{a+c} \right)(a-b)^2 \geq 0$$

or equivalently

$$\sum_{cyc} \frac{(a+c-b)(a-b)^2}{(a+b+c)(a+c)} \geq 0$$

*which is obviously true by the **Triangle inequality**.*

The proof is complete. Equality holds for $a = b = c$.

∎

Solution 4.

We will give four solutions to this problem.

- ***First Solution:***

We have

$$\sum_{cyc} \frac{a(a^2+bc)}{b+c} = \sum_{cyc} \left(\frac{a(a^2+bc)}{b+c} - a^2 \right) + \sum_{cyc} a^2$$

$$= \sum_{cyc} \frac{a(a-b)(a-c)}{b+c} + \sum_{cyc} a^2$$

*Applying **Generalised Schur's inequality**, we get*

$$\sum_{cyc} \frac{a(a-b)(a-c)}{b+c} \geq 0$$

Therefore

$$\sum_{cyc} \frac{a(a^2+bc)}{b+c} \geq \sum_{cyc} a^2$$

*According to **Problem 2**, we have*

$$\sum_{cyc} a^2 \geq 3$$

Finally, we get

$$\sum_{cyc} \frac{a(a^2+bc)}{b+c} \geq 3$$

as desired. Equality holds for $a = b = c = 1$.

- ***Second Solution:***

WLOG, we can assume that $a \geq b \geq c \geq 0$. Therefore

$$a^3 + abc \geq b^3 + abc \geq c^3 + abc$$

$$\frac{1}{b+c} \geq \frac{1}{c+a} \geq \frac{1}{a+b}$$

*Applying **Chebyshev's inequality**, we deduce*

$$\sum_{cyc} \frac{a(a^2+bc)}{b+c} \geq \frac{1}{3}\left(\sum_{cyc}(a^3+abc)\right)\left(\sum_{cyc}\frac{1}{a+b}\right)$$

$$= 2\sum_{cyc}\frac{1}{a+b}$$

*On the other hand, by **AM-GM inequality***

$$\left(\sum_{cyc}(a+b)\right)\left(\sum_{cyc}\frac{1}{a+b}\right) \geq 3\sqrt[3]{\prod_{cyc}(a+b)}.3\sqrt[3]{\prod_{cyc}\left(\frac{1}{a+b}\right)} = 9$$

or equivalently

$$\sum_{cyc}\frac{1}{a+b} \geq \frac{9}{2(a+b+c)}$$

Consequently

$$\sum_{cyc}\frac{a(a^2+bc)}{b+c} \geq 2\sum_{cyc}\frac{1}{a+b}$$

$$\geq \frac{9}{a+b+c}$$

It remains to prove that

$$a+b+c \leq 3$$

*Let $p = a+b+c$, $q = ab+bc+ca$ and $r = abc$. Applying **Schur's inequality**, we obtain*

$$p^3 + 9r \geq 4pq$$

Therefore

$$6 = p^3 - 3pq + 6r$$

$$\geq p^3 - 3pq + \frac{2(4pq - p^3)}{3}$$

$$= \frac{p^3}{3} - \frac{pq}{3}$$

$$\geq \frac{p^3}{3} - \frac{p.p^2}{9}$$

$$= \frac{2p^3}{9}$$

Consequently

$$p \leq 3$$

This ends the proof. Equality holds for $a = b = c = 1$.

● **Third Solution:**

We will prove the following two inequalities

$$\sum_{cyc} \frac{abc}{b+c} \geq ab + bc + ca - \frac{1}{2}(a^2 + b^2 + c^2)$$

$$\sum_{cyc} \frac{a^3}{b+c} \geq \frac{3}{2}(a^2 + b^2 + c^2) - (ab + bc + ca)$$

○ Regarding the first inequality, we have

$$2abc \left(\sum_{cyc} \frac{1}{b+c} \right) + (a^2 + b^2 + c^2) = \sum_{cyc} \left(\frac{2abc}{b+c} + a^2 \right)$$

$$= \sum_{cyc} \frac{a(2bc + ab + ac)}{b+c}$$

$$= \sum_{cyc} \frac{ab(a+c)}{b+c} + \sum_{cyc} \frac{ac(b+a)}{b+c}$$

$$= \sum_{cyc} ab \left(\frac{a+c}{b+c} + \frac{b+c}{a+c} \right)$$

$$\geq \sum_{cyc} ab.2.\sqrt{\frac{a+c}{b+c} \cdot \frac{b+c}{a+c}}$$

$$= 2(ab + bc + ca)$$

○ Regarding the second inequality, we can rewrite it as

$$\sum_{cyc} \left(\frac{2a^3}{b+c} - a^2 \right) \geq \sum_{cyc} (a - b)^2$$

On the other hand, we have

$$\sum_{cyc} \left(\frac{2a^3}{b+c} - a^2 \right) = \sum_{cyc} \frac{a^2(a-b) + a^2(a-c)}{b+c}$$

$$= \sum_{cyc} \frac{(a-b)^2(a^2 + b^2 + ab + bc + ca)}{(b+c)(c+a)}$$

Therefore, we can rewrite the inequality as sum of squares

$$(b-c)^2 S_a + (c-a)^2 S_b + (a-b)^2 S_c \geq 0$$

where

$$\begin{cases} S_a = (b+c)(b^2 + c^2 - a^2) \\ S_b = (c+a)(c^2 + a^2 - b^2) \\ S_c = (a+b)(a^2 + b^2 - c^2) \end{cases}$$

WLOG, we can assume that $a \geq b \geq c$. Since $S_b \geq 0$, $S_c \geq 0$ and

$$S_a + S_b = (a+b)(a-b)^2 + c^2(a+b+2c) \geq 0$$

*According to **SOS theorem**, we deduce*

$$(b-c)^2 S_a + (c-a)^2 S_b + (a-b)^2 S_c \geq 0$$

*Summing up the two inequalities and using **Problem 2**, we get*

$$\sum_{cyc} \frac{a^3+abc}{b+c} = \sum_{cyc} \frac{a^3}{b+c} + abc \left(\sum_{cyc} \frac{1}{b+c} \right)$$

$$\geq \left(\frac{3}{2}(a^2+b^2+c^2) - (ab+bc+ca) \right) + \left(ab+bc+ca - \frac{1}{2}(a^2+b^2+c^2) \right)$$

$$= a^2 + b^2 + c^2$$

$$\geq 3$$

as desired. Equality holds for $a=b=c=1$.

- ***Fourth Solution:***

*We will use **Integrated inequalities** to prove it.*

WLOG, suppose $a \geq b \geq c$. We have

$$\forall x \in [0,1] \qquad a(a-c) \geq b(b-c)x^{a-b}$$

$$\Longleftrightarrow a(a-c)(a-b)x^{b+c-1} \geq b(a-b)(b-c)x^{a+c-1}$$

which gives

$$\sum_{cyc} x^{b+c-1} a(a-b)(a-c) \geq 0$$

that we can rewrite as

$$\sum_{cyc} x^{b+c-1}(a^3+abc) \geq \sum_{cyc} a^2(b+c)x^{b+c-1}$$

Therefore

$$\int_0^1 \sum_{cyc} x^{b+c-1}(a^3+abc) \geq \int_0^1 \sum_{cyc} a^2(b+c)x^{b+c-1}$$

which is equivalent to the desired inequality.

Comment 4.

*We will show in the following example how **Integrated inequalities** could be used to prove some difficult problems.*

- ***Example:** [AoPS]*

Let a, b, c be positive real numbers. Prove that:

$$\frac{1}{4a} + \frac{1}{4b} + \frac{1}{4c} + \frac{1}{a+b} + \frac{1}{b+c} + \frac{1}{c+a} \geq \frac{3}{3a+b} + \frac{3}{3b+c} + \frac{3}{3c+a}$$

○ **Proof:**

Applying **Vasc's inequality**, we obtain

$$(x^2 + y^2 + z^2)^2 \geq 3(x^3 y + y^3 z + z^3 x)$$

Let $x = t^a$, $y = t^b$ and $z = t^c$, thus we get

$$(t^{2a} + y^{2b} + t^{2c})^2 \geq 3(t^{3a+b} + t^{3b+c} + t^{3c+a})$$

Now, integrating this inequality between $[0, 1]$, we get

$$\int_0^1 \frac{(t^{2a} + y^{2b} + t^{2c})^2}{t} \, dt \geq 3 \int_0^1 \frac{(t^{3a+b} + t^{3b+c} + t^{3c+a})}{t} \, dt$$

$$\iff \int_0^1 \left(\sum_{cyc} t^{4a-1} + 2 \sum_{cyc} t^{2a+2b-1} \right) dt \geq 3 \int_0^1 \sum_{cyc} t^{3a+b-1} \, dt$$

which leads to the inequality

$$\frac{1}{4a} + \frac{1}{4b} + \frac{1}{4c} + \frac{1}{a+b} + \frac{1}{b+c} + \frac{1}{c+a} \geq \frac{3}{3a+b} + \frac{3}{3b+c} + \frac{3}{3c+a}$$

∎

Solution 5.

We will give two solutions to this problem.

• First Solution:

According to **Cauchy-Schwarz inequality**, we have

$$(a^3 + b^3 + c^3)(a + b + c) \geq (a^2 + b^2 + c^2)^2$$

which is equivalent to

$$\frac{a^3 + b^3 + c^3}{a^2 + b^2 + c^2} \geq \frac{a^2 + b^2 + c^2}{a+b+c}$$

Therefore, it suffices to prove that

$$\frac{a^2 + b^2 + c^2}{a+b+c} + 1 \geq \frac{2}{3}(a+b+c)$$

Multiplying both sides by $a + b + c$, we obtain

$$(a^2 + b^2 + c^2) + (a+b+c) \geq \frac{2}{3}(a^2 + b^2 + c^2 + 6)$$

which can be rewritten as

$$\frac{1}{3}(a^2 + b^2 + c^2) + (a+b+c) \geq 4$$

We are done immediately since

$$a^2 + b^2 + c^2 \geq ab + bc + ca = 3$$

$$a + b + c = \sqrt{(a+b+c)^2} \geq \sqrt{3(ab+bc+ca)} = 3$$

This ends the proof. Equality holds for $a = b = c = 1$.

- *Second Solution:*

Using uvw notations: $a + b + c = 3u$, $ab + bc + ca = 3v^2$ and $abc = w^3$, we get $v = 1$ and $u \geq 1$.

We can rewrite the inequality as

$$\frac{w^3 + 9u(u^2 - 1)}{3u^2 - 2} + 1 \geq 2u$$

*Applying **Schur's inequality**, we deduce that*

$$w^3 \geq 4uv^2 - 3u^3 = 4u - 3u^3$$

Therefore, it is sufficient to check

$$\frac{4u - 3u^3 + 9u(u^2 - 1)}{3u^2 - 2} + 1 \geq 2u$$

or equivalently

$$\frac{(u-1)(3u+2)}{3u^2 - 2} \geq 0$$

which is obviously true for $u \geq 1$. Equality holds for $a = b = c = 1$.

∎

Solution 6.

*Let $p = a + b + c$, $q = ab + bc + ca$ and $r = abc$. Applying **Schur's inequality**, we get*

$$6 = p^3 - 3pq + 6r$$

$$\geq p^3 - 3pq + \frac{2(4pq - p^3)}{3}$$

$$= \frac{p^3}{3} - \frac{pq}{3}$$

$$\geq \frac{p^3}{3} - \frac{p \cdot p^2}{9}$$

$$= \frac{2p^3}{9}$$

Consequently

$$p \leq 3$$

*On the other hand, by **AM-GM inequality**, we get*

$$pq = (a + b + c)(ab + bc + ca)$$
$$\geq 3\sqrt[3]{abc}.3\sqrt[3]{a^2b^2c^2}$$
$$= 9abc$$
$$= 9r$$

Thus

$$6 = p^3 - 3pq + 6r$$
$$6 \leq p^3 - 27r + 6r$$
$$6 \leq p^3 - 21r$$

Therefore

$$p \geq \sqrt[3]{6}$$

$$r \leq \frac{p^3 - 6}{21}$$

Finally

$$5(a + b + c) - 9 - 6abc \geq 5p - 9 - 6.\frac{p^3 - 6}{21}$$
$$= \frac{(3 - p)(2p^2 + 6p - 17)}{7}$$
$$\geq 0$$

which is true because

$$\sqrt[3]{6} \leq p \leq 3$$

The proof is complete. Equality holds for $a = b = c = 1$.

■

Solution 7.

We will give two solutions to this problem.

• *First Solution:*

*Applying **Cauchy-Schwarz inequality**, we obtain*

$$\frac{a^2 + b^2}{1 + ab} + \frac{b^2 + c^2}{1 + bc} + \frac{c^2 + a^2}{1 + ca} \geq \frac{\left(\sqrt{a^2 + b^2} + \sqrt{b^2 + c^2} + \sqrt{c^2 + a^2}\right)^2}{ab + bc + ca + 3}$$
$$= \frac{2(a^2 + b^2 + c^2) + 2\sum_{cyc}\sqrt{(a^2 + b^2)(a^2 + c^2)}}{ab + bc + ca + 3}$$

On the other hand, we know that

$$(a + b + c)^2 \geq 3(ab + bc + ca)$$

Again by **Cauchy-Schwarz inequality**, we get

$$\sqrt{(a^2 + b^2)(a^2 + c^2)} \geq a^2 + bc$$

Therefore

$$\frac{a^2 + b^2}{1 + ab} + \frac{b^2 + c^2}{1 + bc} + \frac{c^2 + a^2}{1 + ca} \geq \frac{2(a^2 + b^2 + c^2) + 2\sum_{cyc}(a^2 + bc)}{\frac{1}{3}(a + b + c)^2 + 3}$$

$$= \frac{3(a^2 + b^2 + c^2) + (a + b + c)^2}{6}$$

$$= \frac{3(a^2 + b^2 + c^2) + 9}{6}$$

$$\geq \frac{2\sqrt{3(a^2 + b^2 + c^2).9}}{6}$$

$$= \sqrt{3(a^2 + b^2 + c^2)}$$

This ends the proof. Equality holds for $a = b = c = 1$.

- **_Second Solution:_**

We will be using **Theorem 42 (Belabess)**: For $n = 3$ and $a, b, c, x, y, z \geq 0$, we get

$$x(a + b) + y(b + c) + z(c + a) \geq 2\sqrt{(x + y + z)(xab + ybc + zca)}$$

Thus

$$\sum_{cyc} \frac{a^2 + b^2}{1 + ab} = \sum_{cyc} a^2 \left(\frac{1}{1 + ab} + \frac{1}{1 + ac} \right)$$

$$\geq 2\sqrt{(a^2 + b^2 + c^2) \left(\sum_{cyc} \frac{a^2}{(1 + ab)(1 + ac)} \right)}$$

Therefore, it is enough to prove

$$\sum_{cyc} \frac{a^2}{(1 + ab)(1 + ac)} \geq \frac{3}{4}$$

Applying **Cauchy-Schwarz inequality**, we find that

$$\sum_{cyc} \frac{a^2}{(1 + ab)(1 + ac)} \geq \frac{\left(\sum_{cyc} a \right)^2}{\sum_{cyc}(1 + ab)(1 + ac)}$$

$$= \frac{9}{\sum_{cyc}(1 + ab)(1 + ac)}$$

It remains to prove that

$$\sum_{cyc}(1 + ab)(1 + ac) \leq 12$$

On the other hand, we have

$$abc \leq \frac{(a+b+c)^3}{27} = 1$$

$$ab + bc + ca \leq \frac{(a+b+c)^2}{3} = 3$$

Therefore

$$\sum_{cyc}(1+ab)(1+ac) = \sum_{cyc}(1 + ab + ac + a^2bc)$$

$$= 3 + 2(ab + bc + ca) + 3abc$$

$$\leq 12$$

We are done. Equality holds for $a = b = c = 1$.

Comment 5.

*We will use the **Discriminant method** to prove **Theorem 42 (Belabess)** in the particular case for $n = 3$: (The general case could be proved in a similar way).*

$$\sum_{cyc} x(a+b) \geq 2\sqrt{(x+y+z)(xab + ybc + zca)}$$

Let

$$P(t) = x(t-a)(t-b) + y(t-b)(t-c) + z(t-c)(t-a)$$

We have $P(0) \geq 0$ and

$$P(a)P(b)P(c) = -xyz(a-b)^2(b-c)^2(c-a)^2 \leq 0$$

Therefore

$$\Delta = (x(a+b) + y(b+c) + z(c+a))^2 - 4(x+y+z)(xab + ybc + zca) \geq 0$$

as desired.

■

Solution 8.

Rewrite the inequality in the form

$$\frac{1}{a+b} + \frac{1}{b+c} + \frac{1}{c+a} + \frac{abc}{3} = \frac{1}{a+b+c}\left[3 + \sum_{cyc}\frac{a}{b+c}\right] + \frac{abc}{3}$$

*According to **Cauchy-Schwarz inequality***

$$\sum_{cyc}\frac{a}{b+c} \geq \frac{(a+b+c)^2}{2(ab+bc+ca)}$$

Therefore

$$\frac{1}{a+b} + \frac{1}{b+c} + \frac{1}{c+a} + \frac{abc}{3} = \frac{1}{a+b+c}\left[3 + \sum_{cyc}\frac{a}{b+c}\right] + \frac{abc}{3}$$

$$\geq \frac{1}{a+b+c}\left[3+\sum_{cyc}\frac{(a+b+c)^2}{2(ab+bc+ca)}\right]+\frac{abc}{3}$$

$$=\frac{3}{a+b+c}+\frac{(a+b+c)^2}{2(a+b+c)(ab+bc+ca)}+\frac{abc}{3}$$

*Applying **Schur's inequality**, we deduce that*

$$a^3+b^3+c^3+6abc\geq(a+b+c)(ab+bc+ca)$$

Thus

$$\frac{1}{a+b}+\frac{1}{b+c}+\frac{1}{c+a}+\frac{abc}{3}\geq\frac{3}{a+b+c}+\frac{(a+b+c)^2}{2(a+b+c)(ab+bc+ca)}+\frac{abc}{3}$$

$$\geq\frac{3}{a+b+c}+\frac{(a+b+c)^2}{2(a^3+b^3+c^3+6abc)}+\frac{abc}{3}$$

$$=\frac{3}{a+b+c}+\left[\frac{(a+b+c)^2}{6(1+2abc)}+\frac{1+2abc}{6}\right]-\frac{1}{6}$$

$$\geq\frac{3}{a+b+c}+2\sqrt{\frac{(a+b+c)^2}{6(1+2abc)}\cdot\frac{1+2abc}{6}}-\frac{1}{6}$$

$$=\frac{3}{a+b+c}+\frac{a+b+c}{3}-\frac{1}{6}$$

$$\geq 2-\frac{1}{6}$$

$$=\frac{11}{6}$$

as desired. Equality holds for $a=b=c=1$.

∎

Solution 9.

Since $ab+bc+ca+abc=4$, we will first prove that

$$a+b+c\geq 3$$

In fact, we know that

$$(a+b+c)^2\geq 3(ab+bc+ca)$$

*Applying **AM-GM inequality**, we obtain*

$$(a+b+c)^3\geq 27abc$$

Therefore

$$\frac{(a+b+c)^2}{3}+\frac{(a+b+c)^3}{27}\geq 4$$

$$\Longleftrightarrow\frac{(a+b+c-3)(a+b+c+6)^2}{27}\geq 0$$

We conclude that

$$a+b+c\geq 3$$

The original inequality is equivalent to

$$(a + b + c - 1)\left(\frac{1}{a+b} + \frac{1}{b+c} + \frac{1}{c+a}\right) \geq \frac{3(a+b+c+1)}{4}$$

Let $p = a + b + c$, $q = ab + bc + ca$ and $r = abc$. Therefore $p \geq 3$ and $q + r = 4$.

Consider 2 cases:

(i) **The first case: $3 \leq p \leq 4$.**

   *Applying **Schur's inequality**, we obtain*

$$r \geq \frac{p(4q - p^2)}{9} \Rightarrow 4 - q \geq \frac{p(4q - p^2)}{9}$$

   which can be reduced to

$$q \leq \frac{p^3 + 36}{4p + 9}$$

   *According to **Cauchy-Schwarz inequality**, we have*

$$(a + b + c)\left(\frac{1}{a+b} + \frac{1}{b+c} + \frac{1}{c+a}\right) = \frac{a}{b+c} + \frac{b}{c+a} + \frac{c}{a+b} + 3$$

$$\geq \frac{p^2}{2q} + 3$$

$$\geq \frac{4p^3 + 9p^2}{2p^3 + 72} + 3$$

   Therefore

$$\frac{1}{a+b} + \frac{1}{b+c} + \frac{1}{c+a} \geq \frac{4p^2 + 9p}{2p^3 + 72} + \frac{3}{p}$$

   We only need to prove

$$(p - 1)\left(\frac{4p^2 + 9p}{2p^3 + 72} + \frac{3}{p}\right) \geq \frac{3(p+1)}{4}$$

   or equivalently

$$(3 - p)(p - 4)(3p^3 + 4p^2 - 6p + 36) \geq 0$$

   which is true.

(ii) **The second case: $p \geq 4$.**

   *According to **Cauchy-Schwarz inequality**, w obtain*

$$(a + b + c)\left(\frac{1}{a+b} + \frac{1}{b+c} + \frac{1}{c+a}\right) = \frac{a}{b+c} + \frac{b}{c+a} + \frac{c}{a+b} + 3$$

$$\geq \frac{p^2}{2q} + 3$$

$$\geq \frac{p^2}{8} + 3$$

Therefore, we get

$$\frac{1}{a+b} + \frac{1}{b+c} + \frac{1}{c+a} \geq \frac{p}{8} + \frac{3}{p}$$

It remains to prove that

$$(p-1)\left(\frac{p}{8} + \frac{3}{p}\right) \geq \frac{3(p+1)}{4}$$

or equivalently

$$\frac{(p-4)(p^2 - 3p + 6)}{8p} \geq 0$$

which is obviously true for $p \geq 4$.

Equality cases: $(1,1,1)$, $(2,2,0)$ and their permutations.

$\blacksquare$

Solution 10.

We will give three solutions to this inequality.

● *First Solution:*

*From **AM-GM inequality**, we have*

$$6\sqrt{\frac{a^3 + b^3 + c^3}{a+b+c}} \leq a+b+c+ \frac{9(a^3 + b^3 + c^3)}{(a+b+c)^2}$$

Therefore, it is sufficient to prove

$$\frac{(a+b)^2}{c} + \frac{(b+c)^2}{a} + \frac{(c+a)^2}{b} \geq 2(a+b+c) + \frac{18(a^3 + b^3 + c^3)}{(a+b+c)^2}$$

WLOG, we can assume that $a + b + c = 3$. Let

$$f(u) = \frac{(3-u)^2}{u} - 2u^3 - 2$$

The inequality is equivalent to prove

$$f(a) + f(b) + f(c) \geq 3f(1)$$

On the other hand

$$f''(u) = \frac{6(3 - 2u^4)}{u^3}$$

*Thus, f is convex on $]0, 1]$. According to **Vasc's HCF theorem**, it is enough to prove*

$$f(y) + 2f(x) \geq 3f(1)$$

with $y + 2x = 3$ and $x \in]0, \frac{3}{2}[$.

After some simplifications, we get

$$\frac{6(x-1)^2(9 - 24x + 22x^2 - 4x^3)}{x(3 - 2x)} \geq 0$$

which is true for $x \in]0, \frac{3}{2}[$.

- ***Second Solution:***

From **AM-GM inequality**, we have

$$6\sqrt{\frac{a^3 + b^3 + c^3}{a + b + c}} \le a + b + c + \frac{9(a^3 + b^3 + c^3)}{(a + b + c)^2}$$

We need to prove

$$\frac{(a + b)^2}{c} + \frac{(b + c)^2}{a} + \frac{(c + a)^2}{b} \ge 2(a + b + c) + \frac{18(a^3 + b^3 + c^3)}{(a + b + c)^2}$$

that we can transform into

$$\sum_{cyc} \frac{[ab(a - b)^2 + c^2(ab + 8c^2)](a - b)^2}{abc(a + b + c)^2} \ge 0$$

which is obviously true.

- ***Third Solution:***

WLOG, we can assume $a + b + c = 1$. Using uvw notations, we have

$$\frac{(a + b)^2}{c} + \frac{(b + c)^2}{a} + \frac{(c + a)^2}{b} = \frac{3v^2 - 5w^3}{w^3}$$

$$12\sqrt{a^3 + b^3 + c^3} = 12\sqrt{1 - 9v^2 + 3w^3}$$

The inequality is equivalent to prove $f(v^2) \ge 0$ with

$$f(v^2) = \frac{3v^2 - 5w^3}{w^3} - 12\sqrt{1 - 9v^2 + 3w^3}$$

which is an increasing function of v^2. Therefore, by **uvw method**, it is enough to check when v^2 reaches a minimal value which happens when two variables are equal.

WLOG, let $b = a$, hence $c = 1 - 2a$, and we need to prove

$$\frac{2a^2}{1 - 2a} + \frac{(1 - a)^2}{a} \ge 6\sqrt{2a^3 + (1 - 2a)^3}$$

We square both sides to get

$$\frac{(3a - 1)^2(96a^5 - 224a^4 + 152a^3 - 31a^2 - 2a + 1)}{a^2(2a - 1)^2} \ge 0$$

which is true for $a \in]0, \frac{1}{2}[$.

■

Solution 11.

We will give two solutions to this inequality.

- **First Solution:**

*We will use the **Mixing Variables method**. WLOG, we can assume* $c = \min\{a, b, c\}$, *hence* $c \leq 1$.

Let

$$L(a, b, c) = \frac{1}{a+b} + \frac{1}{b+c} + \frac{1}{c+a} + \frac{2(a-1)(b-1)(c-1)}{3} - \frac{3}{2}$$

We have

$$L(a, b, c) - L\left(\frac{a+b}{2}, \frac{a+b}{2}, c\right) = \frac{1}{b+c} + \frac{1}{a+c} - \frac{4}{a+b+2c} + \frac{(1-c)(a-b)^2}{6} \geq 0$$

It is enough to check the case $b = a$ *which gives* $c = 3 - 2a$ *and* $a \leq \frac{3}{2}$. *In this particular case, we get*

$$L(a, a, 3 - 2a) = \frac{(a-1)^2(3-2a)(3+10a-4a^2)}{6(3-a)a} \geq 0$$

which is obviously true. Equality holds for $(1, 1, 1)$, $\left(\frac{3}{2}, \frac{3}{2}, 0\right)$ *and their permutations.*

- **Second Solution:**

Let $t = ab + bc + ca$. *We have*

$$0 \leq t \leq \frac{(a+b+c)^2}{3} = 3$$

(i) **The first case:** $0 \leq t \leq \frac{9}{4}$.

$$
\begin{aligned}
LHS &= \frac{(a+b+c)^2 + (ab+bc+ca)}{(a+b+c)(ab+bc+ca) - abc} + \frac{2abc}{3} - \frac{2(ab+bc+ca)}{3} + \frac{4}{3} \\
&\geq \frac{9+t}{3t} - \frac{2t}{3} + \frac{4}{3} \\
&= \frac{3}{2} + \frac{(9-4t)(2+t)}{6t} \\
&\geq \frac{3}{2}
\end{aligned}
$$

as desired.

(ii) **The second case:** $\frac{9}{4} \leq t \leq 3$.

We have

$$LHS = \frac{1}{a+b+c}\left(3 + \sum_{cyc} \frac{a}{b+c}\right) + \frac{2abc}{3} - \frac{2(ab+bc+ca)}{3} + \frac{4}{3}$$

*Applying **fourth degree Schur's inequality**, we obtain*

$$\sum_{cyc} a^2(a-b)(a-c) \geq 0$$

Rewrite this one in the following form

$$(a^2+b^2+c^2)^2 + 6abc(a+b+c) \geq (a+b+c)^2(ab+bc+ca)$$

Therefore

$$abc \geq \frac{-4t^2 + 45t - 81}{18}$$

*Furthermore, according to **Cauchy-Schwarz inequality**, we get*

$$\sum_{cyc} \frac{a}{b+c} \geq \frac{(a+b+c)^2}{2(ab+bc+ca)} = \frac{9}{2t}$$

Finally

$$LHS \geq \frac{1}{3}\left(3 + \frac{9}{2t}\right) + \frac{-4t^2+45t-81}{27} - \frac{2t}{3} + \frac{4}{3}$$
$$\geq \frac{3}{2} + \frac{(4t-9)(2t-3)(3-t)}{54t}$$
$$\geq \frac{3}{2}$$

as desired.

This ends the proof. Equality holds for $(1,1,1)$, $\left(\frac{3}{2},\frac{3}{2},0\right)$ *and their permutations.*

■

Solution 12.

We will give two solutions to this problem.

• First Solution:

We will use uvw notations: $a+b+c = 3u$, $ab+bc+ca = 3v^2$ *and* $abc = w^3$. *The condition does not depend on u and*

$$4 = ab+bc+ca+abc = 3v^2 + w^3 \leq 3u^2 + u^3$$

which gives

$$u \geq 1$$

We need to prove that $f(u) \geq 0$, *where*

$$f(u) = 3u^2 - 2v^2 - 1 - 2(\sqrt{2}+1)(u-1)$$

But

$$f'(u) = 6u - 2(\sqrt{2}+1) > 0$$

*Therefore, f is an increasing function. Applying **uvw method**, it is enough to prove our inequality for the minimal value of u, which happens when two variables are equal.*

WLOG, let's assume $b = a$. Thus, $a > 0$ and $c = \frac{2-a}{a}$, where $0 < a \leq 2$. We need to prove that

$$(a-1)^2(a-\sqrt{2})^2 \geq 0$$

which is obviously true. Equality holds for $(1,1,1)$, $(\sqrt{2}, \sqrt{2}, \sqrt{2}-1)$ and their permutations.

- *Second Solution:*

We have $ab + bc + ca + abc = 4$, thus there exists (x, y, z) such that $x + y + z = 1$ with

$$\begin{cases} a = \frac{2x}{1-x} \\ b = \frac{2y}{1-y} \\ c = \frac{2z}{1-z} \end{cases}$$

Let

$$f(u) = \frac{4u^2}{(1-u)^2} - 2(\sqrt{2}+1)\frac{2u}{1-u}$$

We can rewrite the inequality as follows

$$f(x) + f(y) + f(z) \geq 3f\left(\frac{1}{3}\right)$$

On the other hand

$$f''(u) = \frac{8((3+\sqrt{2})u - \sqrt{2})}{(1-u)^4}$$

*Therefore, we conclude that f is convex on $[\frac{1}{3}, 1[$ and according to **Vasc's HCF theorem**, it is enough to check the case*

$$f(y) + 2f(x) \geq 3f\left(\frac{1}{3}\right)$$

with $y + 2x = 1$ and $x \in [0, \frac{1}{2}]$.

After some simplifications, we get

$$\frac{(3+2\sqrt{2})(3x-1)^2(x+1-\sqrt{2})^2}{x^4(1-x)^4} \geq 0$$

which is obviously true.

Equality holds for $(1,1,1)$, $(\sqrt{2}, \sqrt{2}, \sqrt{2}-1)$ and their permutations.

■

Solution 13.

We will give two solutions to this problem.

- **First Solution:**

We have $ab + bc + ca + abc = 4$, *thus there exists* (x, y, z) *such that*

$$\begin{cases} a = \frac{2x}{y+z} \\ b = \frac{2y}{z+x} \\ c = \frac{2z}{x+y} \end{cases}$$

We need to prove that

$$\sum_{cyc} \frac{(x + 2y + 2z)(2x - y - z)}{(y+z)(2x + y + z)} \geq 0$$

$$\Longleftrightarrow \sum_{cyc}(x - y)\left(\frac{x + 2y + 2z}{(y+z)(2x + y + z)} - \frac{y + 2z + 2x}{(x+z)(2y + x + z)}\right) \geq 0$$

$$\Longleftrightarrow \sum_{cyc}(x - y)^2(x + y)(x + y + 2z)(x^2 + y^2 + xy - z^2) \geq 0$$

Consequently, it is enough to prove

$$L = \sum_{cyc}(x - y)^2(x + y)(x + y + 2z)(x^2 + y^2 - z^2) \geq 0$$

WLOG, we can assume $x \geq y \geq z$. *Therefore*

$$L \geq (x - z)^2(x + z)(x + z + 2y)(x^2 + z^2 - y^2)$$
$$+ (y - z)^2(y + z)(y + z + 2x)(y^2 + z^2 - x^2)$$
$$\geq (y - z)^2\left((x + z)(x + z + 2y)(x^2 - y^2) + (y + z)(y + z + 2x)(y^2 - x^2)\right)$$
$$= (y - z)^2(x^2 - y^2)^2$$
$$\geq 0$$

as desired. Equality holds for $(1, 1, 1)$, $(2, 2, 0)$ *and their permutations.*

- **Second Solution:**

We have $ab + bc + ca + abc = 4$, *thus there exists* (x, y, z) *such that* $x + y + z = 1$ *with*

$$\begin{cases} a = \frac{2x}{1-x} \\ b = \frac{2y}{1-y} \\ c = \frac{2z}{1-z} \end{cases}$$

Let

$$f(u) = \frac{2(u - 2)(3u - 1)}{u^2 - 1}$$

We can rewrite the inequality as follows

$$f(x) + f(y) + f(z) \geq 3f\left(\frac{1}{3}\right)$$

On the other hand

$$f''(u) = -\frac{24}{(1+u)^3} - \frac{4}{(u-1)^3}$$

Thus

$$f''(u) = 0 \iff u = \frac{\sqrt[3]{6} - 1}{\sqrt[3]{6} + 1} < \frac{1}{3}$$

Therefore, we conclude that f is convex on $[\frac{1}{3}, 1[$ and according to **Vasc's HCF theorem**, it is enough to check the case

$$f(y) + 2f(x) \geq 3f\left(\frac{1}{3}\right)$$

with $y + 2x = 1$ and $x \in [0, \frac{1}{2}]$.

After some simplifications, we get

$$\frac{(1-3x)^2(1-2x)}{x(1-x^2)} \geq 0$$

which is obviously true. Equality holds for $(1,1,1)$, $(2,2,0)$ and their permutations.

∎

Solution 14.

We can rewrite the original inequality in the following form

$$\sum_{cyc} \left(\frac{a^3 + abc}{b^2 + c^2} - a\right) \geq 0$$

or equivalently

$$\sum_{cyc} \frac{a(a-b)(a-c) + ab(a-b) + ac(a-c)}{b^2 + c^2} \geq 0$$

According to **Generalised Schur's inequality**, we get

$$\sum_{cyc} \frac{a(a-b)(a-c)}{b^2 + c^2} \geq 0$$

Therefore, it is sufficient to prove

$$\sum_{cyc} \frac{ab(a-b) + ac(a-c)}{b^2 + c^2} \geq 0$$

which is straightforward as follows

$$\sum_{cyc} \frac{ab(a-b) + ac(a-c)}{b^2 + c^2} = \sum_{cyc} ab(a-b)\left(\frac{1}{b^2 + c^2} - \frac{1}{a^2 + c^2}\right)$$

$$= \sum_{cyc} \frac{ab(a+b)(a-b)^2}{(b^2+c^2)(a^2+c^2)} \geq 0$$

The proof is complete. Equality holds for $a = b = c = 1$.

∎

Solution 15.

We will give two solutions to this problem.

• First Solution:

WLOG, let $c = \max\{a, b, c\}$. After homogenisation, the original inequality becomes

$$(a^3+b^3+c^3)(a+b+c)+8(ab+bc+ca)^2 \geq 4(a^2+b^2+c^2)(ab+bc+ca)+15abc(a+b+c)$$

which is equivalent to

$$(a^3+b^3+c^3)(a+b+c)+16(ab+bc+ca)^2 \geq 4(a+b+c)^2(ab+bc+ca)+15abc(a+b+c)$$

Let $X = (a-b)^2$ and $Y = (a-c)(b-c)$. The inequality can be transformed into

$$(a+b+c)^2(X+Y)+4(c^2X+abY) \geq 4(ab+bc+ca)(X+Y)$$

or equivalently

$$(a^2+b^2+5c^2-2ab-2bc-2ca)X+(a+b-c)^2Y \geq 0$$

which is obviously true because

$$a^2+b^2+5c^2-2(ab+bc+ca) \geq 2a^2+2b^2+2c^2-2(ab+bc+ca)$$
$$= (a-b)^2+(b-c)^2+(c-a)^2$$
$$\geq 0$$

as desired. Equality holds when $a = b = c$, or $a = b$, $a+b = c$.

• Second Solution:

After homogenisation, we need to prove

$$(a^3+b^3+c^3)(a+b+c)+16(ab+bc+ca)^2 \geq 4(a+b+c)^2(ab+bc+ca)+15abc(a+b+c)$$

WLOG, we can assume that $a+b+c = 1$. Using uvw notations, we can rewrite the inequality as follows

$$f(w^3) = 1 - 21v^2 + 144v^4 - 12w^3 \geq 0$$

*f is a decreasing function of w^3. By **uvw method**, it is enough to check when w^3 gets a maximal value which happens when two variables are equal.*

WLOG, let $b = a$. Therefore $c = 1 - 2a$ and the inequality is equivalent to

$$(4a-1)^2(3a-1)^2 \geq 0$$

which is true. Equality holds when $a = b = c$, or $a = b$, $a+b = c$ and permutations.

Solution 16.

We will give four solutions to this problem.

- **First Solution:**

We will be using the following lemma:

○ **Lemma:** *(SOS 5th criterion)*

Let x, y, z be real numbers such that $x + y + z \geq 0$ and $xy + yz + zx \geq 0$. Thus, for all real numbers a, b and c, the following inequality holds

$$(b - c)^2 x + (c - a)^2 y + (a - b)^2 z \geq 0$$

○ **Proof:**

If $z = 0$ then $xy \geq 0$ and $x + y \geq 0$, which gives $x \geq 0$ and $y \geq 0$ and the inequality is obvious.

Thus, we can assume that $xyz \neq 0$. Now, since $x + y + z \geq 0$, we can assume that $y + z \geq 0$. If $y + z = 0$, then $x \geq 0$, $xy \geq 0$ and we obtain $x + y > 0$. Id est, we can assume that $y + z > 0$ and we need to prove that

$$(a - b)^2 z + (a - b + b - c)^2 y + (b - c)^2 x \geq 0$$

$$\Longleftrightarrow (y + z)(a - b)^2 + 2y(a - b)(b - c) + (x + y)(b - c)^2 \geq 0$$

which is a quadratic. Thus, it is enough to check

$$\Delta \leq 0 \Longleftrightarrow y^2 - (y + z)(x + y) \leq 0$$

or equivalently

$$xy + yz + zx \geq 0$$

which is true.

○ **Application:**

It is enough to prove

$$\begin{cases} \sum_{cyc} (4a^2 - 3) \geq 0 \\ \sum_{cyc} (4a^2 - 3)(4b^2 - 3) \geq 0 \end{cases}$$

Let $p = a + b + c$, $q = ab + bc + ca$ and $r = abc$. Using $\textbf{AM-GM inequality}$, we can easily prove $p \geq 3\sqrt{3}$ and $q \geq 9$. Therefore

$$\sum_{cyc} (4a^2 - 3) \geq \frac{4p^2}{3} - 9 \geq 36 - 9 = 27 > 0$$

Now, we are going to prove the following stronger inequality

$$\sum_{cyc} (4a^2 - 3)(4b^2 - 3) \geq 243$$

*Applying **Schur's inequality** on the triplet (ab, bc, ca), we get*

$$q^3 + 9r^2 \geq 4pqr$$

or equivalently

$$r^2 \leq \frac{q^3}{4q - 9}$$

Therefore

$$\sum_{cyc}(4a^2 - 3)(4b^2 - 3) - 243 = 8(2q^2 + 6q - 7r^2 - 27)$$

$$\geq 8(2q^2 + 6q - \frac{7q^3}{4q - 9} - 27)$$

$$= \frac{8(q - 9)(q^2 + 15q - 27)}{4q - 9}$$

$$\geq 0$$

By the above-mentioned lemma, we obtain

$$(4a^2 - 3)(b - c)^2 + (4b^2 - 3)(c - a)^2 + (4c^2 - 3)(a - b)^2 \geq 0$$

The proof is complete.

- ## *Second Solution:*

Using uvw notations, we can rewrite the inequality as

$$f(v^2) = 4v^4 + 3v^2 - 15u^2 \geq 0$$

*Obviously, f is an increasing function of v^2. According to **uvw method**, it is enough to check when v^2 gets a minimal value, which happens when two variables are equal.*

WLOG, let $c = b$. In this case, we get $a + 2b = ab^2$ meaning that $b > 1$ and

$$a = \frac{2b}{b^2 - 1}$$

Therefore

$$\sum_{cyc}(4a^2 - 3)(b - c)^2 = 2(4b^2 - 3)(a - b)^2 \geq 0$$

as desired. Equality holds for $a = b = c = \sqrt{3}$.

- ## *Third Solution:*

After we homogenise the inequality, we need to prove

$$4(ab + bc + ca)(a + b + c) + 9abc \geq \frac{15(a + b + c)^2}{\frac{1}{a} + \frac{1}{b} + \frac{1}{c}}$$

Then, fully expand to arrive at

$$4\left(\sum_{sym} a^3b^2\right) \geq 7\left(\sum_{cyc} a^3bc\right) + \sum_{cyc} a^2b^2c$$

which is true according to **Muirhead's inequality** *as* $(3,2,0) \succ (3,1,1)$ *and* $(3,2,0) \succ (2,2,1)$.

The problem is completely solved. Equality holds for $a = b = c = \sqrt{3}$.

• Fourth Solution:

Let $p = a + b + c$, $q = ab + bc + ca$ *and* $r = abc$. *Using* **AM-GM inequality**, *we can easily prove* $p \geq 3\sqrt{3}$ *and* $q \geq 9$.

Applying **Schur's inequality** *on the triplet* (ab, bc, ca), *we obtain*

$$q^3 + 9r^2 \geq 4pqr$$

or equivalently

$$r^2 \leq \frac{q^3}{4q - 9}$$

Therefore

$$\sum_{cyc}(4a^2 - 3)(b - c)^2 = 2(4q^2 + 9q - 15r^2)$$

$$\geq 2\left(4q^2 + 9q - \frac{15q^3}{4q - 9}\right)$$

$$= \frac{2q(q-9)(q+9)}{4q - 9}$$

$$\geq 0$$

as desired. Equality holds for $a = b = c = \sqrt{3}$.

■

Solution 17.

We will give two solutions to this inequality.

• First Solution:

The inequality is equivalent to

$$2(a + b + c)^2 \sum_{cyc} a^2(b - c)^2 \geq abc\left[3(a + b + c)^3 - 81abc\right]$$

Let $x = b + c$, $y = c + a$ *and* $z = a + b$. *The inequality becomes*

$$8(x + y + z)\sum_{cyc}(y + z - x)^2(y - z)^2 \geq 3\prod_{cyc}(x + y - z)\left[(x + y + z)^3 - 27\prod_{cyc}(x + y - z)\right]$$

Using sRr notations, the inequality is equivalent to

$$256s^2 r^2 (16R^2 + 8Rr - 3s^2 + r^2) \geq 192s^2 r^2 (s^2 - 27r^2)$$
$$\Longleftrightarrow 64R^2 + 32Rr + 85r^2 \geq 15s^2$$

*Applying **Blundon's inequality**, we only need to prove*

$$64R^2 + 32Rr + 85r^2 \geq 15 \left[2R^2 + 10Rr - r^2 + 2\sqrt{R(R - 2r)^3} \right]$$

$$\Longleftrightarrow 34R^2 - 118Rr + 100r^2 \geq 30\sqrt{R(R - 2r)^3}$$
$$\Longleftrightarrow 4(R - 2r)^2 (8R - 25r)^2 \geq 0$$

which is obvious.

Comment 6.

*We will show here how to prove **Blundon's inequality**.*

Let σ_1, σ_2 and σ_3 be the elementary symmetric functions of the sides a, b, c of a triangle. We first compute σ_1, σ_2 and σ_3 in terms of s, r and R. We have

$$\sigma_1 = a + b + c = 2s$$

and

$$\sigma_3 = abc = 4srR$$

*To compute σ_2, we use **Heron's formula***

$$A = \sqrt{s(s - a)(s - b)(s - c)}$$

where A is the area of the triangle with side lengths a, b, c. Since $A = sr$, we have that

$$r^2 = \frac{(s - a)(s - b)(s - c)}{s}$$
$$= \frac{s^3 - s^2 (a + b + c) + s(ab + bc + ca) - abc}{s}$$
$$= -s^2 + \sigma_2 - 4Rr$$

Consequently

$$\sigma_2 = s^2 + r^2 + 4Rr$$

Consider the symmetric polynomial $(a - b)^2 (b - c)^2 (c - a)^2$ of a, b, c. We know that it can be written as a polynomial of σ_1, σ_2 and σ_3, and hence as a polynomial of s, r, R. More precisely, one checks easily that

$$(a - b)^2 (b - c)^2 (c - a)^2 = \sigma_1^2 \sigma_2^2 - 4\sigma_1^3 \sigma_3 + 18\sigma_1 \sigma_2 \sigma_3 - 27\sigma_3^2$$
$$= -4r^2 [(s^2 - 2R^2 - 10Rr + r^2)^2 - 4R(R - 2r)^3]$$

Therefore

$$(s^2 - 2R^2 - 10Rr + r^2)^2 \leq 4R(R - 2r)^3$$

as desired.

- **Second Solution:**

The case $abc = 0$ is trivial. So let's assume $abc > 0$.

Let $a + b + c = 3u$, $ab + bc + ca = 3v^2$ and $abc = w^3$, and we need to prove that $f(w^3) \geq 0$, where

$$f(w^3) = w^6 - 5u^3 w^3 + 4u^2 v^4$$

but

$$f'(w^3) = 2w^3 - 5u^3 \leq 0$$

f is a decreasing function. Applying **uvw method**, it is enough to prove our inequality for a maximal value of w^3, which happens for equality case of two variables. Since the inequality $w^6 - 5u^3 w^3 + 4u^2 v^4 \geq 0$ is homogeneous, we can assume that $b = c = 1$, which gives

$$(a - 1)^2 (a - 4)^2 \geq 0,$$

which is obviously true.

■

Solution 18.

Let $a + b + c = 3u$, $ab + bc + ca = 3v^2$ and $abc = w^3$. Thus, $3u^2 - 2v^2 = 1$ and we need to prove that

$$3(27u^3 - 27uv^2 + 3w^3) + 2w^3 \geq 11\sqrt{(3u^2 - 2v^2)^3},$$

which is a linear inequality of w^3. By **uvw method**, it is enough to prove our inequality for the extreme values of w^3, which happens in the following cases:

(i) **The first case:** $w^3 = 0$.

Let $c = 0$ and $a^2 + b^2 = 2tab$. Thus, $t \geq 1$ and we need to prove that

$$3(a^3 + b^3) \geq 11\sqrt{\frac{(a^2 + b^2)^3}{27}}$$

or

$$243(a + b)^2 (a^2 - ab + b^2)^2 \geq 121(a^2 + b^2)^3$$

or

$$243(t + 1)(2t - 1)^2 \geq 484t^3$$

or

$$(t - 1)(488t^2 + 488t - 241) + 2 \geq 0$$

which is true for $t \geq 1$.

(ii) **The second case:** two variables are equal.

Let $b = c = 1$ in the homogeneous form. We need to prove that

$$3a^3 + 2a + 6 \geq 11\sqrt{\frac{(a^2 + 2)^3}{27}}$$

After squaring both sides, we get

$$(a-1)^2(61a^4 + 122a^3 - 18a^2 + 328a + 2) \geq 0$$

which is obviously true.

∎

Solution 19.

○ *The left inequality:*

*According to **Cauchy-Schwarz inequality**, we obtain*

$$\sum_{cyc} \sqrt{a^2 - ab + b^2} = \sqrt{\sum_{cyc}\left(2a^2 - ab + 2\sqrt{(a^2 - ab + b^2)(a^2 - ac + c^2)}\right)}$$

$$= \sqrt{\sum_{cyc}\left(2a^2 - ab + 2\sqrt{\left(\left(a - \frac{b}{2}\right)^2 + \frac{3}{4}b^2\right)\left(\left(a - \frac{c}{2}\right)^2 + \frac{3}{4}c^2\right)}\right)}$$

$$\geq \sqrt{\sum_{cyc}\left(2a^2 - ab + 2\left(\left(a - \frac{b}{2}\right)\left(a - \frac{c}{2}\right) + \frac{3}{4}bc\right)\right)}$$

$$= \sqrt{\sum_{cyc}(4a^2 - ab)}$$

$$\geq \sqrt{3(a^2 + b^2 + c^2)}$$

$$= 3$$

The problem is completely solved. Equality holds for $a = b = c = 1$.

○ *The right inequality:*

WLOG, let's assume $c = \min\{a, b, c\}$. We have

$$\sum_{cyc} \sqrt{a^2 - ab + b^2} \leq \sqrt{a^2 - ab + b^2} + a + b$$

$$\leq \sqrt{(1+2)\left(a^2 - ab + b^2 + \frac{(a+b)^2}{2}\right)}$$

$$= 3\sqrt{\frac{a^2 + b^2}{2}}$$

$$\leq 3\sqrt{\frac{a^2 + b^2 + c^2}{2}}$$

$$= 3\sqrt{\frac{3}{2}}$$

This ends the proof. Equality holds for $\left(\frac{3}{2}, \frac{3}{2}, 0\right)$ and permutations.

Solution 20.

*Applying **Mildorf's inequaity** with $k = 3$, we obtain*

$$\sqrt[3]{\frac{a^3 + b^3}{2}} \leq \frac{a^2 + b^2}{a + b}$$

Therefore, it is sufficient to prove

$$\frac{a + b}{a^2 + b^2} + \frac{b + c}{b^2 + c^2} + \frac{c + a}{c^2 + a^2} \geq \frac{3(a + b + c)}{a^2 + b^2 + c^2}$$

which we can do as follows

$$(a^2 + b^2 + c^2)\left(\sum_{cyc} \frac{b + c}{b^2 + c^2}\right) - 3(a + b + c) = \sum_{cyc}\left(\frac{a^2(b + c)}{b^2 + c^2} - a\right)$$

$$= \sum_{cyc} \frac{ab(a - b) + ac(a - c)}{b^2 + c^2}$$

$$= \sum_{cyc} ab(a - b)\left(\frac{1}{b^2 + c^2} - \frac{1}{c^2 + a^2}\right)$$

$$= \sum_{cyc} \frac{ab(a + b)(a - b)^2}{(c^2 + a^2)(c^2 + b^2)}$$

$$\geq 0$$

The proof is complete. Equality holds for $a = b = c$.

■

Solution 21.

We will give four solutions to this inequality.

- **First Solution:**

We will use the following lemma:

○ **Lemma:**

For all non-negative real numbers x, y and z, we have

$$x^2 + y^2 + z^2 + 2xyz + 1 \geq 2(xy + yz + zx)$$

○ **Proof:**

Let

$$L(x, y, z) = x^2 + y^2 + z^2 + 2xyz + 1 - 2(xy + yz + zx)$$

*By **Dirichlet's box principle**: of the three numbers $x - 1$, $y - 1$, $z - 1$, at least two are of the same sign. WLOG, let's assume that $(y - 1)(z - 1) \geq 0$. Therefore*

$$L(x, y, z) = x^2 + y^2 + z^2 + 2xyz + 1 - 2(xy + yz + zx)$$

$$= (x-1)^2 + (y-z)^2 + 2x + 2xyz - 2(xy + zx)$$
$$= (x-1)^2 + (y-z)^2 + 2x(y-1)(z-1)$$
$$\geq 0$$

Equality occurs for $x = y = z = 1$.

○ ***Application:***

Applying the above-mentioned lemma, we obtain

$$(a+2)(b+2)(c+2) - 3(a+b+c)^2 = abc + 4(a+b+c-ab-bc-ac) - 1$$
$$\geq ab + bc + ca - 2 + 4(a+b+c-ab-bc-ac) - 1$$
$$= 4(a+b+c) - 3(ab+bc+ca) - 3$$
$$= (3 - (a+b+c))(3(a+b+c) + 1)$$
$$\geq \left(3 - \sqrt{3(a^2+b^2+c^2)}\right)(3(a+b+c) + 1)$$
$$= 0$$

as required. Equality holds for $a = b = c = 1$.

● ***Second Solution:***

*We will be using the **Mixing Variables method**. The inequality is equivalent to*

$$abc - 4\left(\sum_{cyc} ab\right) + 4\left(\sum_{cyc} a\right) - 1 \geq 0$$

Let

$$f(a,b,c) = abc - 4(ab+bc+ca) + 4(a+b+c) - 1$$

Let $t = \frac{b^2+c^2}{2}$ *and consider*

$$f(a,t,t) = a\left(\frac{b^2+c^2}{2}\right) - 4\left(a\sqrt{2(b^2+c^2)} + \frac{b^2+c^2}{2}\right) + 4\left(a + \sqrt{2(b^2+c^2)}\right) - 1$$

Observe that

$$f(a,b,c) - f(a,t,t) = -\frac{a(b-c)^2}{2} + 2(b-c)^2 + (4a-4)(\sqrt{2(b^2+c^2)} - b - c)$$
$$= (b-c)^2\left(-\frac{a}{2} + 2 + \frac{4a-4}{\sqrt{2(b^2+c^2)} + b + c}\right)$$

If $a \in [1, \sqrt{3}]$ *then the line is obviously positive.*

But if $a \in (0,1]$, *then*

$$\sqrt{2(b^2+c^2)} + b + c \geq 3$$

Thus

$$-\frac{a}{2} + 2 + \frac{4a-4}{\sqrt{2(b^2+c^2)} + b + c} \geq -\frac{1}{2} + 2 - \frac{4}{3} > 0$$

Consequently

$$f(a,b,c) \geq f(a,t,t) \geq \cdots \geq f(1,1,1) = 0$$

and the inequality is proved. Equality holds for $a = b = c = 1$.

● **Third Solution:**

The inequality is equivalent to prove

$$bc(4-a) + 4(b+c)(a-1) \leq 4a - 1$$

Suppose $a = \max\{a,b,c\}$. We have $a \geq 1$ and

$$b + c \leq \frac{a^2 + bc}{a}$$

Therefore, it suffices to prove that

$$bc(4-a) + \frac{4(a-1)(a^2+bc)}{a} \leq 4a - 1$$

$$\Longleftrightarrow bc(8a - a^2 - 4) + 4a^3 - 8a^2 + a \leq 0$$

$$\Longleftrightarrow \frac{b^2+c^2}{2}(8a - a^2 - 4) + 4a^3 - 8a^2 + a \leq 0$$

which is equivalent to

$$(a-1)^2(a^2 + 2a - 12) \leq 0$$

This last inequality is obviously true. Equality holds for $a = b = c = 1$.

● **Fourth Solution:**

WLOG, we can assume $(a - 1)(b - 1) \leq 0$ and the inequality is equivalent to prove

$$(a+2)(b+2)(c+2) - 3(a+b+c)^2 = (c-4)(a-1)(b-1) + 3(1 - c(a+b-1))$$

On the one hand

$$(c-4)(a-1)(b-1) \geq 0$$

Consequently, it is enough to prove

$$1 - c(a+b-1) \geq 0$$

Let $t = \sqrt{a^2 + b^2}$. Therefore

$$c(a+b-1) \leq \frac{c^2+1}{2}\left(\sqrt{2(a^2+b^2)} - 1\right)$$

$$= \frac{4-t^2}{2}\left(\sqrt{2}t - 1\right)$$

$$= 1 - \frac{(3+\sqrt{2}t)(t-\sqrt{2})^2}{2}$$

$$\leq 1$$

This ends the proof. Equality holds for $a = b = c = 1$.

Comment 7.

*We will show in the next example how **Dirichlet's box principle** could be used in solving inequalities.*

○ **Example:** *[APMO 2004]*

Let a, b, c be non-negative real numbers. Prove that:

$$(a^2 + 2)(b^2 + 2)(c^2 + 2) \geq 9(ab + bc + ca)$$

○ **Proof:**

*By **Dirichlet's box principle:** of the three numbers $a^2 - 1$, $b^2 - 1$, $c^2 - 1$, at least two are of the same sign. WLOG, let's assume that $(a^2 - 1)(b^2 - 1) \geq 0$. Therefore*

$$a^2 b^2 + 1 \geq a^2 + b^2$$

Thus

$$(a^2 + 2)(b^2 + 2) = a^2 b^2 + 4 + 2(a^2 + b^2)$$
$$\geq 3(a^2 + b^2 + 1)$$

*Using **Cauchy-Schwarz inequality**, we obtain*

$$(a^2 + 2)(b^2 + 2)(c^2 + 2) \geq 3(a^2 + b^2 + 1)(c^2 + 2)$$
$$= 3(a^2 + b^2 + 1)(1 + 1 + c^2)$$
$$\geq 3(a + b + c)^2$$

Now, it is enough to prove

$$(a + b + c)^2 \geq 3(ab + bc + c)$$

which is trivial. Equality holds for $a = b = c = 1$.

■

Solution 22.

We will give two solutions to this problem.

• **First Solution:**

Let $3u = a + b + c$, $3v^2 = ab + bc + ca$ and $w^3 = abc$. We need to prove $f(w^3) \geq 0$ where

$$f(w^3) = 3w^6 + (14u^3 - 18uv^2)w^3 + 6u^6 - 8u^4 v^2 - 9u^2 v^4 + 12v^6$$

On the other hand

$$f'(w^3) = 6w^3 + 14u^3 - 18uv^2$$
$$= 4(3u^3 - 4uv^2 + w^3) + 2u(u^2 - v^2) + 2w^3$$
$$\geq 0$$

*Since $f(w^3)$ is increasing, by **uvw method**, it is enough to check when w^3 reaches a minimum value.*

(i) The first case: $b = c$.

This case is just **Schur's inequality.**

(ii) The second case: $a = 0$.

$$(6 - b^2 c^2)(b - c)^2 \geq \left(6 - \left(\frac{b+c}{2}\right)^4\right)(b - c)^2$$

$$= \frac{15}{16}(b - c)^2$$

$$\geq 0$$

as desired.

The proof is complete.

- **Second Solution:**

Using the **Buffalo Way technique**, we are going to prove the following stronger inequality

$$a(a - b)(a - c) + b(b - c)(b - a) + c(c - a)(c - b) \geq \frac{16}{27}(a - b)^2(b - c)^2(c - a)^2$$

WLOG, we can assume $a = \min\{a, b, c\}$ and let $x = b - a \geq 0$ and $y = c - a \geq 0$. Therefore, we have $x + y = 3(1 - a) \leq 3$ and we can rewrite the inequality as follows

$$a(x^2 - xy + y^2) + (x - y)^2(x + y) \geq \frac{16}{27}x^2 y^2 (x - y)^2$$

Consequently, it is enough to prove

$$x + y \geq \frac{16}{27}x^2 y^2$$

which is immediate by **AM-GM inequality**

$$\frac{16}{27}x^2 y^2 \leq \frac{16}{27}\left(\frac{(x + y)^2}{4}\right)^2$$

$$= \frac{(x + y)^4}{27}$$

$$\leq x + y$$

We are done. Equality holds for $(1, 1, 1)$, $(0, \frac{3}{2}, \frac{3}{2})$ and their permutations.

Comment 8.

We will show in the following examples how the **Buffalo Way technique** could be used to prove some inequalities.

○ *First example:*

Prove that for any x, y, z non-negative real numbers, we have:

$$\sum_{cyc}(x^3 - x^2y - x^2z + xyz) \geq 0$$

○ *Proof:*

Let $x = \min\{x, y, z\}$, $y = x + u$ and $z = x + v$. We have $u, v \geq 0$ and we can rewrite the inequality as follows

$$\sum_{cyc}(x^3 - x^2y - x^2z + xyz) = (u^2 - uv + v^2)x + (u + v)(u - v)^2 \geq 0$$

which is obviously true.

○ *Second example:*

Prove that for any a, b, c non-negative real numbers, we have:

$$4(a + b + c)^3 \geq 27\left(a^2b + b^2c + c^2a + abc\right)$$

○ *Proof:*

Since the inequality is homogeneous, we can assume that $a = \min\{a, b, c\} = 1$, and let $b = 1 + u$ and $c = 1 + v$.

Therefore

$$4(a + b + c)^3 - 27(a^2b + b^2c + c^2a + abc) = (4u + v)(u - 2v)^2 + 9(u^2 - uv + v^2) \geq 0$$

which is obviously true.

■

Solution 23.

We will give two solutions for this problem.

● *First Solution:*

*Using **Cauchy-Schwarz inequality**, we obtain*

$$\sum_{cyc}\frac{a^3}{a^2 + bc} \geq \frac{(a^2 + b^2 + c^2)^2}{a^3 + b^3 + c^3 + 3abc}$$

*According to **Problem 2**, we obtain*

$$4(a^2 + b^2 + c^2)^3 \geq 3(a^3 + b^3 + c^3 + 3abc)^2$$

Therefore

$$\sum_{cyc} \frac{a^3}{a^2 + bc} \geq \frac{(a^2 + b^2 + c^2)^2}{a^3 + b^3 + c^3 + 3abc}$$

$$\geq \frac{(a^2 + b^2 + c^2)^2}{\sqrt{\frac{4(a^2+b^2+c^2)^3}{3}}}$$

$$= \frac{\sqrt{3(a^2 + b^2 + c^2)}}{2}$$

$$= \frac{3}{2}$$

This ends the proof. Equality holds for $a = b = c = 1$.

● *Second Solution:*

*Applying **Hölder's inequality**, we obtain*

$$\left(\sum_{cyc} \frac{a^3}{a^2 + bc} \right)^2 \geq \frac{(a^2 + b^2 + c^2)^3}{\sum_{cyc} (a^2 + bc)^2}$$

On the other hand

$$\sum_{cyc}(a^2 + bc)^2 = \sum_{cyc}(a^4 + 2a^2bc + b^2c^2)$$

$$\leq (a^2 + b^2 + c^2)^2 + \frac{(a^2 + b^2 + c^2)^2}{3}$$

$$= \frac{4(a^2 + b^2 + c^2)^2}{3}$$

Therefore

$$\left(\sum_{cyc} \frac{a^3}{a^2 + bc} \right)^2 \geq \frac{3(a^2 + b^2 + c^2)^3}{4(a^2 + b^2 + c^2)^2}$$

$$= \frac{3(a^2 + b^2 + c^2)}{4}$$

$$= \frac{9}{4}$$

as desired. Equality holds for $a = b = c = 1$.

Comment 9.

Hölder's inequality *is a powerful technique in proving inequalities, especially those with fractions and radicals. The inequality is very effective and easy to use, but it is often neglected by students. We will show in the following examples how it can be used to prove some difficult problems.*

○ **First example:** [JBMO 2002]

Let a, b, c be positive real numbers. Prove that:

$$\frac{1}{a(a+b)} + \frac{1}{b(b+c)} + \frac{1}{c(c+a)} \geq \frac{27}{2(a+b+c)^2}$$

○ **Proof:**

According to **Hölder's inequality**, we get

$$2\left(\sum_{cyc} a\right)^2 \left(\sum_{cyc} \frac{1}{a(a+b)}\right) = \left(\sum_{cyc} a\right)\left(\sum_{cyc}(a+b)\right)\left(\sum_{cyc} \frac{1}{a(a+b)}\right)$$

$$\geq \left(\sum_{cyc} \sqrt[3]{\frac{a(a+b)}{a(a+b)}}\right)^3$$

$$= 27$$

as desired. Equality holds for $a = b = c$.

○ **Second example:** [IMO 2001]

Let a, b, c be positive real numbers. Prove that:

$$\frac{a}{\sqrt{a^2 + 8bc}} + \frac{b}{\sqrt{b^2 + 8ca}} + \frac{c}{\sqrt{c^2 + 8ab}} \geq 1$$

○ **Proof:**

According to **Hölder's inequality**, we get

$$\left(\sum_{cyc} \frac{a}{\sqrt{a^2 + 8bc}}\right)^2 \left(\sum_{cyc} a(a^2 + 8bc)\right) \geq (a + b + c)^3$$

Therefore, it is sufficient to prove

$$(a + b + c)^3 \geq \sum_{cyc} a(a^2 + 8bc)$$

or equivalently

$$c(a-b)^2 + a(b-c)^2 + b(c-a)^2 \geq 0$$

which is obviously true. Equality holds for $a = b = c$.

○ **Third example:** [AoPS]

Let a, b, c be positive real numbers such that $a + b + c = abc$. Prove that:

$$a^7(bc - 1) + b^7(ca - 1) + c^7(ab - 1) \geq 162\sqrt{3}$$

○ *Proof:*

First of all, we have

$$a^7(bc-1) + b^7(ca-1) + c^7(ab-1) = a^6(abc-a) + b^6(abc-b) + c^6(abc-c)$$
$$= a^6(b+c) + b^6(c+a) + c^6(a+b)$$

*According to **Hölder's inequality**, we get*

$$a^7(bc-1) + b^7(ca-1) + c^7(ab-1) = \sum_{cyc} a^6(b+c)$$
$$\geq \frac{\left(\sum_{cyc} a(b+c)\right)^6}{\left(\sum_{cyc}(b+c)\right)^5}$$
$$= 2\frac{(ab+bc+ca)^6}{(a+b+c)^5}$$

On the other hand, we know that

$$(ab+bc+ca)^2 \geq 3abc(a+b+c)$$

Therefore

$$a^7(bc-1) + b^7(ca-1) + c^7(ab-1) \geq 54(a+b+c)$$

*According to **AM-GM inequality**, we obtain*

$$(a+b+c)^3 \geq 27abc = 27(a+b+c)$$

Consequently

$$a+b+c \geq 3\sqrt{3}$$

Finally, we deduce

$$a^7(bc-1) + b^7(ca-1) + c^7(ab-1) \geq 54(a+b+c)$$
$$\geq 162\sqrt{3}$$

The proof is complete. Equality holds for $a = b = c = \sqrt{3}$.

■

Solution 24.

We will give three solutions to this problem.

● ***First Solution:***

We will use the following famous inequality

$$a^2 + b^2 + c^2 + 2abc + 1 \geq 2(ab+bc+ca)$$

Let

$$L = 2\sqrt{1 + \sum_{cyc} a^2 - \sum_{cyc} ab + abc}$$

The inequality is equivalent to prove

$$L \ge a + b + c$$

We have

$$L^2 = 4\left(1 + \sum_{cyc} a^2 - \sum_{cyc} ab\right) + (abc)^2 + 4abc\sqrt{1 + \sum_{cyc} a^2 - \sum_{cyc} ab}$$

$$\ge 4\left(1 + \sum_{cyc} a^2 - \sum_{cyc} ab\right) + (abc)^2 + 4abc$$

$$= 3 + 4\left(\sum_{cyc} a^2 - \sum_{cyc} ab\right) + \left(1 + (abc)^2\right) + 4abc$$

$$\ge 3 + 4\left(\sum_{cyc} a^2 - \sum_{cyc} ab\right) + 6abc$$

$$= \sum_{cyc} a^2 - 4\left(\sum_{cyc} ab\right) + 3\left(\sum_{cyc} a^2 + 2abc + 1\right)$$

On the other hand, we know that

$$\sum_{cyc} a^2 + 2abc + 1 \ge 2\left(\sum_{cyc} ab\right)$$

Therefore

$$L^2 \ge \sum_{cyc} a^2 - 4\left(\sum_{cyc} ab\right) + 3\left(\sum_{cyc} a^2 + 2abc + 1\right)$$

$$\ge \sum_{cyc} a^2 - 4\left(\sum_{cyc} ab\right) + 6\left(\sum_{cyc} ab\right)$$

$$= (a + b + c)^2$$

Finally

$$\sqrt{1 + \sum_{cyc} a^2 - \sum_{cyc} ab} \ge \frac{a + b + c - abc}{2}$$

The problem is completely solved. Equality holds for $a = b = c = 1$.

• **_Second Solution:_**

If $a + b + c - abc \le 2$, then the inequality is obvious.

Consider the case $a + b + c \geq abc + 2$: according to **AM-GM** and **Schur's inequalities**

$$a^2 + b^2 + c^2 + 2abc + 1 \geq a^2 + b^2 + c^2 + 3(abc)^{2/3}$$
$$\geq a^2 + b^2 + c^2 + \frac{9abc}{a+b+c}$$
$$\geq 2(ab + bc + ca)$$

It follows that

$$1 + a^2 + b^2 + c^2 - ab - bc - ca \geq \frac{(a+b+c)^2 - 6abc + 1}{4}$$

Now, let $x = a + b + c$ and $y = abc$ $(x \geq y + 2)$. We need to prove that

$$\sqrt{x^2 - 6y + 1} \geq x - y$$
$$\Longleftrightarrow 2xy + 1 \geq y^2 + 6y$$

which is true since

$$2xy + 1 \geq 2y(y+2) + 1 = y^2 + 6y + (y-1)^2 \geq y^2 + 6y$$

The proof is complete. Equality holds for $a = b = c = 1$.

• **Third Solution:**

Let $a + b + c = 3u$, $ab + bc + ca = 3v^2$ and $abc = w^3$, and we need to prove

$$2\sqrt{1 + 9u^2 - 9v^2} + w^3 \geq 3u$$

Applying **uvw method**, it is enough to prove our inequality for the minimal value of w^3, which happens in the following cases:

1. **The first case:** $w^3 = 0$.

    Let $c = 0$, and we need to prove that

    $$2\sqrt{1 + a^2 - ab + b^2} \geq a + b$$

    that can be written as

    $$4 + 3(a - b)^2 \geq 0$$

    which is obviously true.

2. **The second case:** two variables are equal.

    Let $b = a$, and we need to prove that

    $$2\sqrt{1 + (a - c)^2} + (a^2 - 1)c \geq 2a$$

    The case $2a - (a^2 - 1)c < 0$ is obvious. Let's assume $2a - (a^2 - 1)c \geq 0$.

    There are 2 cases to consider:

(i) Case: $a^2 \geq 3$.

It's enough to prove
$$2\sqrt{1+(a-c)^2} + 2c \geq 2a$$
$$\Longleftrightarrow 2\sqrt{1+(a-c)^2} \geq 2(a-c)$$

which is obvious.

(ii) Case: $a^2 < 3$.

We need to prove that
$$4(1+(a-c)^2) \geq (2a-(a^2-1)c)^2$$
$$\Longleftrightarrow (3-a^2)(a^2+1)c^2 - 4a(3-a^2)c + 4 \geq 0$$

which is a quadratic in c. Therefore, it is enough to check that $\Delta \leq 0$
$$\Delta = a^2(3-a^2)^2 - (3-a^2)(1+a^2) \leq 0$$
$$\Longleftrightarrow \Delta = -(3-a^2)(a^2-1)^2 \leq 0$$

which is true.

The proof is complete.

∎

Solution 25.

We will give two solutions to this inequality.

• *First Solution:*

We will prove the following stronger inequality
$$\sqrt{1+\frac{1}{2}(a-b)^2} + \sqrt{1+\frac{1}{2}(b-c)^2} + \sqrt{1+\frac{1}{2}(c-a)^2} \geq a+b+c$$

After squaring both sides, we get
$$3 + \sum_{cyc} \frac{(a-b)^2}{2} + 2\sum_{cyc} \sqrt{\left(1+\frac{(a-b)^2}{2}\right)\left(1+\frac{(a-c)^2}{2}\right)} \geq (a+b+c)^2$$

Using **Cauchy-Schwarz inequality**, we get
$$\sqrt{\left(1+\frac{(a-b)^2}{2}\right)\left(1+\frac{(a-c)^2}{2}\right)} \geq 1+\frac{(a-b)(a-c)}{2}$$

Therefore, it is sufficient to prove
$$3 + \sum_{cyc} \frac{(a-b)^2}{2} + 2\sum_{cyc} \left(1+\frac{(a-b)(a-c)}{2}\right) \geq (a+b+c)^2$$

which can be simplified into

$$a^2 + b^2 + c^2 + 9 \geq 4(ab + bc + ca)$$

or equivalently

$$(a + b + c)^2 + 9 \geq 6(ab + bc + ca)$$

On the other hand $a + b + c = ab + bc + ca$, so we need to prove

$$(a + b + c)^2 + 9 \geq 6(a + b + c)$$

which is equivalent to

$$(a + b + c - 3)^2 \geq 0$$

The proof is complete. Equality holds for $a = b = c = 1$.

• **Second Solution:**

Let $f : x \to \sqrt{1 + x^2}$. We have

$$f''(x) = \frac{1}{(1 + x^2)^{\frac{3}{2}}}$$

*Therefore, f is a convex function. Applying **Jensen's inequality**, we obtain*

$$\sum_{cyc} \sqrt{1 + (a - b)^2} \geq \sqrt{9 + (|a - b| + |b - c| + |c - a|)^2}$$

$$\geq \sqrt{9 + 2\left((a - b)^2 + (b - c)^2 + (c - a)^2\right)}$$

$$= \sqrt{9\left(\frac{ab + bc + ca}{a + b + c}\right)^2 + 2\left((a - b)^2 + (b - c)^2 + (c - a)^2\right)}$$

Thus, it suffices to prove that

$$9\left(\frac{ab + bc + ca}{a + b + c}\right)^2 + 2((a - b)^2 + (b - c)^2 + (c - a)^2) \geq (a + b + c)^2$$

which is equivalent to

$$\sum_{cyc}(a^4 + b^2 c^2) \geq 2\sum_{cyc} a^2 bc$$

*This last inequality is true by **Muirhead's inequality** as $(4, 0, 0) \succ (2, 1, 1)$ and $(2, 2, 0) \succ (2, 1, 1)$.*

Comment 10.

Jensen's inequality *is probably the most powerful technique in solving inequalities but unfortunately finding the right weights could be very challenging. We will show in the following examples how a good choice of weights could help in solving some difficult problems.*

○ **First example:** *[AoPS]*

Let a, b, c be positive real numbers. Prove that:

$$\sqrt{\frac{a}{a+b}} + \sqrt{\frac{b}{b+c}} + \sqrt{\frac{c}{c+a}} \le \frac{3\sqrt{2}}{2}$$

○ **Proof:**

Let $f : x \to \sqrt{x}$. We can rewrite the inequality as

$$\sqrt{\frac{a}{a+b}} + \sqrt{\frac{b}{b+c}} + \sqrt{\frac{c}{c+a}} = \sum_{cyc} \frac{a+c}{2(a+b+c)} \cdot \sqrt{\frac{4a(a+b+c)^2}{(a+b)(a+c)^2}}$$

$$= \sum_{cyc} \frac{a+c}{2(a+b+c)} f\left(\frac{4a(a+b+c)^2}{(a+b)(a+c)^2}\right)$$

We also have

$$\sum_{cyc} \frac{a+c}{2(a+b+c)} = 1$$

f is a concave function, thus by **Jensen's inequality**

$$\sum_{cyc} \frac{a+c}{2(a+b+c)} f\left(\frac{4a(a+b+c)^2}{(a+b)(a+c)^2}\right) \le f\left(\sum_{cyc} \frac{a+c}{2(a+b+c)} \cdot \frac{4a(a+b+c)^2}{(a+b)(a+c)^2}\right)$$

$$= \sqrt{\sum_{cyc} \frac{2a(a+b+c)}{(a+b)(a+c)}}$$

Therefore, it is sufficient to prove

$$\frac{2a(a+b+c)}{(a+b)(a+c)} + \frac{2b(a+b+c)}{(b+c)(b+a)} + \frac{2c(a+b+c)}{(c+a)(c+b)} \le \frac{9}{2}$$

which is equivalent to

$$\sum_{cyclic} a^2 b \ge 6abc$$

This last inequality is true by **AM-GM inequality**. The proof is complete.

○ **Second example:** *[AoPS]*

Let a, b, c be positive real numbers. Prove that:

$$\frac{a^2}{a+b} + \frac{b^2}{b+c} + \frac{c^2}{c+a} \ge \frac{a+b+c}{2}$$

○ **Proof:**

Let $f : x \to \frac{1}{1+x}$. We can rewrite the inequality as

$$\frac{a^2}{a+b} + \frac{b^2}{b+c} + \frac{c^2}{c+a} = (a+b+c) \sum_{cyc} \frac{a}{a+b+c} \cdot \frac{a}{a+b}$$

$$= (a+b+c) \sum_{cyc} \frac{a}{a+b+c} \cdot \frac{1}{1 + \frac{b}{a}}$$

$$= (a+b+c) \sum_{cyc} \frac{a}{a+b+c} \cdot f\left(\frac{b}{a}\right)$$

f is a convex function on $[0, +\infty[$, *so by **Jensen's inequality***

$$\frac{a^2}{a+b} + \frac{b^2}{b+c} + \frac{c^2}{c+a} = (a+b+c) \sum_{cyc} \frac{a}{a+b+c} \cdot f\left(\frac{b}{a}\right)$$

$$\geq (a+b+c).f\left(\sum_{cyc} \frac{a}{a+b+c} \cdot \frac{b}{a}\right)$$

$$= (a+b+c).f(1)$$

$$= \frac{a+b+c}{2}$$

as desired. Equality at $a = b = c$.

○ ***Third example:*** *[HSGS]*

Let a, b, c *be positive real numbers. Prove that:*

$$\frac{a^2}{\sqrt{3a^2 + 8b^2 + 14ab}} + \frac{b^2}{\sqrt{3b^2 + 8c^2 + 14bc}} + \frac{c^2}{\sqrt{3c^2 + 8a^2 + 14ca}} \geq \frac{a+b+c}{5}$$

○ ***Proof:***

Let $f : x \to \frac{1}{\sqrt{3+8x^2+14x}}$. *We can rewrite the inequality as*

$$\sum_{cyc} \frac{a^2}{\sqrt{3a^2 + 8b^2 + 14ab}} = (a+b+c) \sum_{cyc} \frac{a}{a+b+c} \cdot \frac{a}{\sqrt{3a^2 + 8b^2 + 14ab}}$$

$$= (a+b+c) \sum_{cyc} \frac{a}{a+b+c} \cdot \frac{1}{\sqrt{3 + 8\left(\frac{b}{a}\right)^2 + 14\left(\frac{b}{a}\right)}}$$

$$= (a+b+c) \sum_{cyc} \frac{a}{a+b+c} \cdot f\left(\frac{b}{a}\right)$$

On the other hand, we have

$$f''(x) = \frac{128x^2 + 224x + 123}{(8x^2 + 14x + 3)^{\frac{5}{2}}}$$

Therefore f is a convex function on $[0, +\infty[$, *so by **Jensen's inequality***

$$\sum_{cyc} \frac{a^2}{\sqrt{3a^2 + 8b^2 + 14ab}} = (a+b+c) \sum_{cyc} \frac{a}{a+b+c} \cdot f\left(\frac{b}{a}\right)$$

$$\geq (a+b+c).f\left(\sum_{cyc} \frac{a}{a+b+c} \cdot \frac{b}{a}\right)$$

$$= (a+b+c).f(1)$$
$$= \frac{a+b+c}{5}$$

as desired. Equality at $a = b = c$.

∎

Solution 26.

We are going to prove the following stronger result

$$\sum_{cyc} \frac{(a+b-c)^2}{(a+b)^2+c^2} \geq \frac{3(a^2+b^2+c^2)}{a^2+b^2+c^2+4(ab+bc+ca)}$$

Using **Cauchy-Schwarz inequality**, we have

$$\sum_{cyc} \frac{(b+c-a)^2}{a^2+(b+c)^2} \geq \frac{\left[\sum_{cyc}(b+c-a)^2\right]^2}{\sum_{cyc}(b+c-a)^2[a^2+(b+c)^2]}$$

Therefore, it is enough to check

$$\left[\sum_{cyc}(b+c-a)^2\right]^2\left(\sum_{cyc}a^2+4\sum_{cyc}bc\right) \geq 3\left(\sum_{cyc}a^2\right)\left(\sum_{cyc}(b+c-a)^2[a^2+(b+c)^2]\right)$$

Notice that

$$\left[\sum_{cyc}(b+c-a)^2\right]^2 = 9\left(\sum_{cyc}a^2\right)^2 - 4\left(\sum_{cyc}bc\right)\left(3\sum_{cyc}a^2 - \sum_{cyc}bc\right)$$

Furthermore

$$\sum_{cyc}(b+c-a)^2[a^2+(b+c)^2] = 3\left(\sum_{cyc}a^2\right)^2 + 4\left(\sum_{cyc}bc\right)^2 - 16abc\left(\sum_{cyc}a\right)$$

Therefore, the above inequality is equivalent to

$$9\left(\sum_{cyc}a^2\right)^2\left(\sum_{cyc}a^2+4\sum_{cyc}bc\right) - 3\left(\sum_{cyc}a^2\right)\left[3\left(\sum_{cyc}a^2\right)^2 + 4\left(\sum_{cyc}bc\right)^2 - 16abc\left(\sum_{cyc}a\right)\right]$$
$$- 4\left(\sum_{cyc}bc\right)\left(3\sum_{cyc}a^2 - \sum_{cyc}bc\right)\left(\sum_{cyc}a^2+4\sum_{cyc}bc\right) \geq 0$$

which can be transformed into

$$3\left(\sum_{cyc}a^2\right)\left(\sum_{cyc}bc\right)\left(3\sum_{cyc}a^2 - \sum_{cyc}bc\right) - \left(\sum_{cyc}bc\right)\left(3\sum_{cyc}a^2 - \sum_{cyc}bc\right)\left(\sum_{cyc}a^2+4\sum_{cyc}bc\right)$$
$$+ 12abc\left(\sum_{cyc}a\right)\left(\sum_{cyc}a^2\right) \geq 0$$

97

Consequently, it is enough to prove

$$\left(\sum_{cyc} bc\right)\left(3\sum_{cyc} a^2 - \sum_{cyc} bc\right)\left(\sum_{cyc} a^2 - 2\sum_{cyc} bc\right) + 6abc\left(\sum_{cyc} a\right)\left(\sum_{cyc} a^2\right) \geq 0$$

*Using **fourth-degree Schur's inequality**, we obtain*

$$\sum_{cyc} a^2 - 2\sum_{cyc} bc \geq \frac{-6abc\left(\sum_{cyc} a\right)}{\sum_{cyc} a^2 + \sum_{cyc} bc}$$

It remains to prove that

$$-6abc\left(\sum_{cyc} a\right)\left(\sum_{cyc} bc\right)\left(3\sum_{cyc} a^2 - \sum_{cyc} bc\right) + 6abc\left(\sum_{cyc} a\right)\left(\sum_{cyc} a^2\right)\left(\sum_{cyc} a^2 + \sum_{cyc} bc\right) \geq 0$$

which reduces to the obvious inequality

$$6abc\left(\sum_{cyc} a\right)\left(\sum_{cyc} a^2 - \sum_{cyc} bc\right)^2 \geq 0$$

which is obviously true.

Equality holds if and only if $a = b = c$, or $a = b, c = 0$, or $a > 0, b = c = 0$ and their permutations.

∎

Solution 27.

We will give two solutions to this problem.

● ***First Solution:***

We need to prove

$$\sum_{cyc} \frac{a^2}{bc} + \sum_{cyc} \frac{bc}{a^2} \geq 2(a^3 + b^3 + c^3)$$

*Applying **AM-GM inequality**, we obtain*

$$LHS \geq 2\sqrt{\left(\sum_{cyc} \frac{a^2}{bc}\right)\left(\sum_{cyc} \frac{bc}{a^2}\right)} = 2\sqrt{(a^3 + b^3 + c^3)\left(\frac{1}{a^3} + \frac{1}{b^3} + \frac{1}{c^3}\right)}$$

Thus, it suffices to prove that

$$\frac{1}{a^3} + \frac{1}{b^3} + \frac{1}{c^3} \geq a^3 + b^3 + c^3$$

that can be transformed into

$$\frac{1}{a^3} + (b^3 + c^3)\left(\frac{1}{b^3 c^3} - 1\right) \geq a^3$$

WLOG, we can assume that $a \geq b \geq c$. Therefore $a \geq 1$ and we get

$$\begin{cases} bc \leq \frac{(b+c)^2}{4} = \frac{(3-a)^2}{4} \leq 1 \\ b^3 + c^3 \geq \frac{(b+c)^3}{4} = \frac{(3-a)^3}{4} \end{cases}$$

Consequently, it is sufficient to prove

$$\frac{1}{a^3} + \frac{(3-a)^3}{4}\left(\frac{64}{(3-a)^6} - 1\right) \geq a^3$$

or equivalently

$$\frac{3(a-1)^2(a^7 - 4a^6 - 18a^5 + 112a^4 - 163a^3 + 48a^2 + 36a + 36)}{4(3-a)^3 a^3} \geq 0$$

which is true for $1 \leq a < 3$.

Equality holds for $a = b = c = 1$.

- **_Second Solution:_**

Let: $3u = a + b + c$, $3v^2 = ab + bc + ca$ *and* $w^3 = abc$. *We need to prove* $f(w^3) \leq 0$ *where*

$$f(w^3) = 2w^9 + (16u^3 - 18uv^2)w^6 - (9u^6 - 18u^4v^2)w^3 - 9u^3v^6$$

On the other hand, we have

$$\begin{aligned} f''(w^3) &= 12w^3 + 32u^3 - 36uv^2 \\ &= 10(3u^3 - 4uv^2 + w^3) + 2u^3 + 4uv^2 + 2w^3 \\ &\geq 0 \end{aligned}$$

*Since $f(w^3)$ is convex, by **uvw method**, it is enough to check when w^3 reaches critical value, which happens when two variables are equal.*

WLOG, we can assume $b = c = 1$. In this case, we obtain

$$(a-1)^2(a^5 + 10a^4 - 11a^3 + 2a^2 + 44a + 8) \geq 0$$

which is true.

Equality holds for $a = b = c = 1$.

■

Solution 28.

According to **AM-GM inequality,** *we have*

$$\frac{a^2}{b^2+c^2} = \frac{a^3}{a(b^2+c^2)}$$

$$\geq \frac{a^3}{\frac{a^3+2b^3}{3} + \frac{a^3+2c^3}{3}}$$

$$= \frac{3a^3}{2(a^3+b^3+c^3)}$$

Therefore

$$\sqrt[3]{\frac{a^2}{b^2+c^2}} \geq \sqrt[3]{\frac{3a^3}{2(a^3+b^3+c^3)}}$$

$$= \frac{\sqrt[3]{3}a}{\sqrt[3]{2(a^3+b^3+c^3)}}$$

Similarly

$$\sqrt[3]{\frac{b^2}{c^2+a^2}} \geq \frac{\sqrt[3]{3}b}{\sqrt[3]{2(a^3+b^3+c^3)}}$$

$$\sqrt[3]{\frac{c^2}{a^2+b^2}} \geq \frac{\sqrt[3]{3}c}{\sqrt[3]{2(a^3+b^3+c^3)}}$$

Summing up all these inequalities, we obtain

$$\sqrt[3]{\frac{a^2}{b^2+c^2}} + \sqrt[3]{\frac{b^2}{c^2+a^2}} + \sqrt[3]{\frac{c^2}{a^2+b^2}} \geq \frac{\sqrt[3]{3}(a+b+c)}{\sqrt[3]{2(a^3+b^3+c^3)}}$$

The proof is complete. Equality holds for $a = b = c$.

Comment 11.

The motivation behind this solution is the **Isolated Fudging technique** *where the main idea is to compare the individual terms to expressions such as*

$$\frac{a^r}{a^r+b^r+c^r} \quad or \quad \frac{a^r+b^r}{a^r+b^r+c^r}$$

We will show in the following examples a systematic way to find the value of r.

○ *First example: [AoPS]*

Let a, b, c be non-negative real numbers. Prove that:

$$\frac{1+ab+ac}{(1+b+c)^2} + \frac{1+bc+ba}{(1+c+a)^2} + \frac{1+ca+cb}{(1+a+b)^2} \geq 1$$

○ **Proof:**

Using the **Isolated Fudging technique**, we need to find r such that

$$\frac{1 + ab + ac}{(1 + b + c)^2} \geq \frac{a^r}{a^r + b^r + c^r}$$

Let

$$f(a, b, c) = \frac{1 + ab + ac}{(1 + b + c)^2} - \frac{a^r}{a^r + b^r + c^r}$$

The point $(1, 1, 1)$ should be a local minimum and as such we need

$$\frac{\partial f}{\partial a}(1, 1, 1) = 0$$

Calculating the partial derivative, we get

$$\frac{\partial f}{\partial a} = \frac{b + c}{(b + c + 1)^2} - \frac{r a^{r-1}(a^r + b^r + c^r) - a^r.r.a^{r-1}}{(a^r + b^r + c^r)^2}$$

Evaluating the partial derivative at $(a, b, c) = (1, 1, 1)$, we get

$$\frac{2}{9} - \frac{3r - r}{9} = 0$$

which means that

$$r = 1$$

Now that we have found a candidate for r, we only need to prove

$$\frac{1 + ab + ac}{(1 + b + c)^2} \geq \frac{a}{a + b + c}$$

which can be done as follows

$$\frac{1 + ab + ac}{(1 + b + c)^2} - \frac{a}{a + b + c} = \frac{(a - 1)^2(b + c)}{(a + b + c)(1 + b + c)^2} \geq 0$$

(*Alternatively:* $(1 + ab + ac)(1 + \frac{b}{a} + \frac{c}{a}) \geq (1 + b + c)^2$)

Similarly, we have

$$\frac{1 + bc + ba}{(1 + c + a)^2} \geq \frac{b}{a + b + c}$$

$$\frac{1 + ca + cb}{(1 + a + b)^2} \geq \frac{c}{a + b + c}$$

Summing up all these inequalities, we get

$$\frac{1 + ab + ac}{(1 + b + c)^2} + \frac{1 + bc + ba}{(1 + c + a)^2} + \frac{1 + ca + cb}{(1 + a + b)^2} \geq 1$$

The proof is complete. Equality holds for $a = b = c = 1$.

○ **Second example:** *[IMO 2001]*

Let a, b, c be positive real numbers. Prove that:

$$\frac{a}{\sqrt{a^2 + 8bc}} + \frac{b}{\sqrt{b^2 + 8ca}} + \frac{c}{\sqrt{c^2 + 8ab}} \geq 1$$

○ **Proof:**

*Using the **Isolated Fudging technique**, we need to find r such that*

$$\frac{a}{\sqrt{a^2 + 8bc}} \geq \frac{a^r}{a^r + b^r + c^r}$$

Let

$$f(a, b, c) = \frac{a}{\sqrt{a^2 + 8bc}} - \frac{a^r}{a^r + b^r + c^r}$$

The point $(1, 1, 1)$ should be a local minimum and a such we need

$$\frac{\partial f}{\partial a}(1, 1, 1) = 0$$

Now, calculating the partial derivative

$$\frac{\partial f}{\partial a} = \frac{\sqrt{a^2 + 8bc} - \frac{a^2}{\sqrt{a^2 + 8bc}}}{a^2 + 8bc} - \frac{ra^{r-1}(a^r + b^r + c^r) - a^r . r . a^{r-1}}{(a^r + b^r + c^r)^2}$$

Evaluating the partial derivative at $(a, b, c) = (1, 1, 1)$, we get

$$\frac{3 - \frac{1}{3}}{9} - \frac{3r - r}{9} = 0$$

which means that

$$r = \frac{4}{3}$$

Now that we have found r, we only need to prove

$$\frac{a}{\sqrt{a^2 + 8bc}} \geq \frac{a^{\frac{4}{3}}}{a^{\frac{4}{3}} + b^{\frac{4}{3}} + c^{\frac{4}{3}}}$$

*According to **AM-GM inequality**, we get*

$$\left(a^{\frac{4}{3}} + b^{\frac{4}{3}} + c^{\frac{4}{3}}\right)^2 = (a^{\frac{4}{3}})^2 + (b^{\frac{4}{3}} + c^{\frac{4}{3}})(a^{\frac{4}{3}} + a^{\frac{4}{3}} + b^{\frac{4}{3}} + c^{\frac{4}{3}})$$

$$\geq (a^{\frac{4}{3}})^2 + 2b^{\frac{2}{3}}c^{\frac{2}{3}} . 4a^{\frac{2}{3}}b^{\frac{1}{3}}c^{\frac{1}{3}}$$

$$= a^{\frac{8}{3}} + 8a^{\frac{2}{3}}bc$$

$$= a^{\frac{2}{3}}(a^2 + 8bc)$$

Therefore

$$\frac{a}{\sqrt{a^2 + 8bc}} \geq \frac{a^{\frac{4}{3}}}{a^{\frac{4}{3}} + b^{\frac{4}{3}} + c^{\frac{4}{3}}}$$

Similarly, we get

$$\frac{b}{\sqrt{b^2 + 8ca}} \geq \frac{b^{\frac{4}{3}}}{a^{\frac{4}{3}} + b^{\frac{4}{3}} + c^{\frac{4}{3}}}$$

$$\frac{c}{\sqrt{c^2 + 8ab}} \geq \frac{c^{\frac{4}{3}}}{a^{\frac{4}{3}} + b^{\frac{4}{3}} + c^{\frac{4}{3}}}$$

Summing up all these inequalities, we get the desired result. Equality holds for $a = b = c$.

○ ***Third example:*** *[AoPS]*

Let a, b, c be non-negative real numbers with positive sum. Prove that:

$$\sqrt{\frac{a^3}{a^3 + (b+c)^3}} + \sqrt{\frac{b^3}{b^3 + (c+a)^3}} + \sqrt{\frac{c^3}{c^3 + (a+b)^3}} \geq 1$$

○ ***Proof:***

*Using the **Isolated Fudging technique**, we need to find r such that*

$$\sqrt{\frac{a^3}{a^3 + (b+c)^3}} \geq \frac{a^r}{a^r + b^r + c^r}$$

Let

$$f(a, b, c) = \sqrt{\frac{a^3}{a^3 + (b+c)^3}} - \frac{a^r}{a^r + b^r + c^r}$$

The point $(1, 1, 1)$ should be a local minimum and as such we need

$$\frac{\partial f}{\partial a}(1, 1, 1) = 0$$

Calculating the partial derivative, we get

$$\frac{\partial f}{\partial a} = \frac{\frac{3a^2}{a^3 + (b+c)^3} - \frac{3a^5}{(a^3 + (b+c)^3)^2}}{2\sqrt{\frac{a^3}{a^3 + (b+c)^3}}} - \frac{ra^{r-1}(a^r + b^r + c^r) - a^r . r . a^{r-1}}{(a^r + b^r + c^r)^2}$$

Evaluating the partial derivative at $(a, b, c) = (1, 1, 1)$, we obtain

$$\frac{4}{9} - \frac{3r - r}{9} = 0$$

which means

$$r = 2$$

Now that we have found our candidate for r, it is enough to prove

$$\sqrt{\frac{a^3}{a^3 + (b+c)^3}} \geq \frac{a^2}{a^2 + b^2 + c^2}$$

Let $t = \frac{b+c}{a}$. *We have*

$$\sqrt{\frac{a^3}{a^3 + (b+c)^3}} = \frac{1}{\sqrt{1+t^3}}$$

$$= \frac{1}{\sqrt{(1+t)(1-t+t^2)}}$$

$$\geq \frac{1}{\frac{1+t}{2} + \frac{1-t+t^2}{2}}$$

$$= \frac{1}{1 + \frac{t^2}{2}}$$

$$\geq \frac{1}{1 + \frac{b^2 + c^2}{a^2}}$$

$$= \frac{a^2}{a^2 + b^2 + c^2}$$

Similarly

$$\sqrt{\frac{b^3}{b^3 + (c+a)^3}} \geq \frac{b^2}{a^2 + b^2 + c^2}$$

$$\sqrt{\frac{c^3}{c^3 + (a+b)^3}} \geq \frac{c^2}{a^2 + b^2 + c^2}$$

Summing up all these inequalities, we get

$$\sqrt{\frac{a^3}{a^3 + (b+c)^3}} + \sqrt{\frac{b^3}{b^3 + (c+a)^3}} + \sqrt{\frac{c^3}{c^3 + (a+b)^3}} \geq 1$$

The proof is complete. Equality holds for a = b = c.

○ **Fourth example:** *[MOP 2002]*

Let a, b, c *be positive real numbers. Prove that:*

$$\left(\frac{2a}{b+c}\right)^{\frac{2}{3}} + \left(\frac{2b}{c+a}\right)^{\frac{2}{3}} + \left(\frac{2c}{a+b}\right)^{\frac{2}{3}} \geq 3$$

○ **Proof:**

*Using the **Isolated Fudging technique**, we need to find r such that*

$$\left(\frac{2a}{b+c}\right)^{\frac{2}{3}} \geq \frac{3a^r}{a^r + b^r + c^r}$$

Let

$$f(a, b, c) = \left(\frac{2a}{b+c}\right)^{\frac{2}{3}} - \frac{3a^r}{a^r + b^r + c^r}$$

104

The point $(1,1,1)$ should be a local minimum and as such we need

$$\frac{\partial f}{\partial a}(1,1,1) = 0$$

Calculating the partial derivative, we get

$$\frac{\partial f}{\partial a} = \frac{\sqrt[3]{32}}{3\sqrt[3]{a(b+c)^2}} - \frac{3(ra^{r-1}(a^r + b^r + c^r) - a^r.r.a^{r-1})}{(a^r + b^r + c^r)^2}$$

Evaluating the partial derivative at $(a,b,c) = (1,1,1)$, we get the following equation

$$\frac{2}{3} - \frac{3r - r}{3} = 0$$

which means

$$r = 1$$

Therefore, it is sufficient to prove

$$\left(\frac{2a}{b+c}\right)^{\frac{2}{3}} \geq \frac{3a}{a+b+c}$$

which can be proved using **AM-GM inequality** as follows

$$\frac{3a}{a+b+c} \leq \frac{3a}{\sqrt[3]{a\left(\frac{b+c}{2}\right)^2}}$$

$$= \left(\frac{2a}{b+c}\right)^{\frac{2}{3}}$$

Similarly

$$\left(\frac{2b}{c+a}\right)^{\frac{2}{3}} \geq \frac{3b}{a+b+c}$$

$$\left(\frac{2c}{a+b}\right)^{\frac{2}{3}} \geq \frac{3c}{a+b+c}$$

Summing up all these inequalities, we get

$$\left(\frac{2a}{b+c}\right)^{\frac{2}{3}} + \left(\frac{2b}{c+a}\right)^{\frac{2}{3}} + \left(\frac{2c}{a+b}\right)^{\frac{2}{3}} \geq 3$$

The proof is complete. Equality holds for $a = b = c$.

■

Solution 29.

We will give two solutions to this problem.

● *First Solution:*

We will use uvw notations: $a + b + c = 3u$, $ab + bc + ca = 3v^2$ *and* $abc = w^3$. *We need to prove*

$$\frac{54u^3 - 63uv^2 + 9w^3}{9uv^2 - w^3} \geq \frac{8(9u^4 - 12u^2v^2 + 2v^4 + uw^3)}{9u^4}$$

or

$$f(w^3) \geq 0$$

where

$$f(w^3) = 486u^7 - 1215u^2v^2 + 864u^3v^4 - 144uv^6 + 153u^4w^3 - 168u^2v^2w^3 + 16v^4w^3 + 8uw^6$$

Now, by **Schur's inequality**, *we obtain*

$$f'(w^3) = 16uw^3 + 153u^4 - 168u^2v^2 + 16v^4$$
$$\geq 16u(4uv^2 - 3u^3) + 153u^4 - 168u^2v^2 + 16v^4$$
$$= 105u^4 - 104u^2v^2 + 16v^4$$
$$> 0$$

f is an increasing function. Applying **uvw method**, *it is enough to prove our inequality for the minimal value of* w^3, *which happens in the following cases:*

(i) *The first case:* $c \to 0^+$ *and* $b = 1$.

We need to prove that

$$a + \frac{1}{a} \geq \frac{3}{2} + \frac{4(a^4 + 1)}{(a + 1)^4}$$

which can be transformed to an obviousness

$$(a - 1)^2(2a^4 + a^3 + 2a^2 + a + 2) \geq 0$$

(ii) *The second case:* $b = c = 1$.

In this case, we get

$$a^4(a - 1)^2 \geq 0$$

This ends the proof.

● *Second Solution:*

We have

$$\text{LHS-RHS} = \frac{(a + b + c)\sum_{cyc} a^4(a - b)(a - c) + \sum_{cyc} a^5(a - b)(a - c)}{(a + b)(b + c)(c + a)(a + b + c)^4} \geq 0$$

which is true by **sixth and seventh degree Schur's inequalities**.

Solution 30.

We will give two solutions to this problem.

- *First Solution:*

WLOG, we can assume that $a \geq b \geq c$. Therefore $a + b \geq 2$, and we get

$$2(a + b) \geq (a + c)(b + c)$$

On the other hand, we have

$$\frac{bc + a}{b + c} + \frac{ca + b}{c + a} - 2 = \frac{(2 - c)(a - b)^2 + c(a + b)(a + b - 2)}{2(b + c)(a + c)} + \frac{c^2(a + b - 2)}{2(a + c)(b + c)}$$

$$\geq \frac{(2 - c)(a - b)^2 + c(a + b)(a + b - 2)}{2(b + c)(a + c)}$$

$$\geq \frac{(2 - c)(a - b)^2 + c(a + b)(a + b - 2)}{4(a + b)}$$

$$\geq \frac{(a - b)^2 + c(a + b)^2 - 2c(a + b)}{4(a + b)}$$

$$\geq \frac{(a - b)^2 + 4c(a + b) - 4c - 2c(a + b)}{4(a + b)}$$

$$\geq \frac{(a - b)^2 + c(a + b) - 4c}{4(a + b)}$$

$$= \frac{3}{4} - \frac{ab + c}{a + b}$$

The inequality is proved. Equality holds for $\left(\frac{3}{2}, \frac{3}{2}, 0\right)$ and permutations.

- *Second Solution:*

We need to prove that

$$\sum_{cyc} \frac{3ab + c(a + b + c)}{a + b} \geq \frac{11}{4}(a + b + c)$$

or equivalently

$$\sum_{cyc} \frac{c^2 + 3ab}{a + b} \geq \frac{7}{4}(a + b + c)$$

After some calculations, we get

$$\sum_{cyc}(4a^4 - 3a^3b - 3a^3c - 2a^2b^2 + 12a^2bc) \geq 0$$

or

$$4\sum_{cyc}(a^4 - a^3b - a^3c + a^2bc) + \sum_{cyc}(a^3b + a^3c - 2a^2b^2) + 8abc(a + b + c) \geq 0$$

which can be transformed into

$$4\sum_{cyc} a^2(a-b)(a-c) + \sum_{cyc} ab(a-b)^2 + 8abc(a+b+c) \geq 0$$

*This last inequality is true by **fourth degree Schur's inequality**.*

∎

Solution 31.

*Applying **Titu's lemma**, we obtain*

$$\frac{\sum_{cyc} a^2}{\sum_{cyc} ab} + \frac{1}{2}\sum_{cyc} \frac{ab}{a^2+b^2} = \frac{a^2+b^2+c^2}{ab+bc+ca} + \frac{1}{4}\sum_{cyc}\frac{(a+b)^2}{a^2+b^2} - \frac{3}{4}$$

$$\geq \frac{a^2+b^2+c^2}{ab+bc+ca} + \frac{(a+b+c)^2}{2(a^2+b^2+c^2)} - \frac{3}{4}$$

$$= \frac{a^2+b^2+c^2}{ab+bc+ca} + \frac{ab+bc+ca}{a^2+b^2+c^2} - \frac{1}{4}$$

$$\geq 2 - \frac{1}{4}$$

$$= \frac{7}{4}$$

as desired. Equality holds for $a=b=c$.

∎

Solution 32.

We will give two solutions to this problem.

- ***First Solution:***

Using pqr notations: $p = a+b+c = 3$, $q = ab+bc+ca$ and $r = abc$, we can rewrite the inequality as follows

$$f(r) = \frac{(9+q)^2 - 12(3q-r)}{(3q-r)^2} + \frac{r}{4} \geq 1$$

There are 2 cases to consider:

(i) ***The first case:*** $q \leq \frac{9}{4}$.

*Using **Iran 1996's inequality**, we get*

$$LHS \geq \sum_{cyc} \frac{1}{(a+b)^2} \geq \frac{9}{4q} \geq 1$$

(*ii*) **The second case:** $\frac{9}{4} \leq q \leq 3$.

According to **Schur's inequality**, we get

$$r \geq \frac{\left(4q - p^2\right)\left(p^2 - q\right)}{6p} = \frac{(4q - 9)(9 - q)}{18}$$

On the other hand, f is an increasing function

$$f\left(r\right) \geq f\left(\frac{(4q - 9)(9 - q)}{18}\right)$$

$$= \frac{108\left(36q + 81 - 5q^2\right)}{\left(4q^2 + 9q + 81\right)^2} + \frac{(4q - 9)(9 - q)}{72}$$

Therefore, it is sufficient to prove

$$\frac{108\left(36q + 81 - 5q^2\right)}{\left(4q^2 + 9q + 81\right)^2} + \frac{(4q - 9)(9 - q)}{72} \geq 1$$

$$\Longleftrightarrow (q - 3)(4q - 9)\left(16q^4 - 24q^3 + 297q^2 - 2268q + 13851\right) \leq 0$$

which is obviously true.

This ends the proof. Equality holds for $(1, 1, 1)$, $(\frac{3}{2}, \frac{3}{2}, 0)$ and their permutations.

- **Second Solution:**

Using *uvw* notations: $a + b + c = 3u$, $ab + bc + ca = 3v^2$ and $abc = w^3$, we can rewrite the original inequality as

$$f(w^3) = \frac{(9 + 3v^2)^2 - 108v^2 + 12w^3}{(9v^2 - w^3)^2} + \frac{w^3}{4} - 1 \geq 0$$

f is an increasing function and by **uvw method**, we only need to check 2 cases:

(*i*) **The first case:** $w^3 \to 0^+$.

WLOG, we can assume that $c \to 0^+$. We have $a + b = 3$ and we need to prove

$$\frac{1}{a^2} + \frac{1}{b^2} \geq \frac{8}{9}$$

Applying **AM-GM inequality**, we deduce

$$\frac{1}{a^2} + \frac{1}{b^2} \geq \frac{1}{2}\left(\frac{1}{a} + \frac{1}{b}\right)^2$$

$$\geq \frac{1}{2}\frac{4^2}{(a + b)^2}$$

$$= \frac{8}{9}$$

(ii) **The second case:** $b = a$.

We have $c = 3 - 2a$ and $a \leq \frac{3}{2}$. In this particular case, we get

$$\frac{(a-1)^2(3-2a)(a^4 - 4a^3 + 6a + 3)}{4a^2(a-3)^2} \geq 0$$

which is true.

The proof is complete. Equality holds for $(1,1,1)$, $\left(\frac{3}{2}, \frac{3}{2}, 0\right)$ and their permutations.

Comment 12.

We present here three different proofs of **Iran 1996's inequality**.

○ **First proof:**

Let $a = x + y$, $b = y + z$ and $c = z + x$. We need to prove

$$\left(2ab + 2bc + 2ca - a^2 - b^2 - c^2\right)\left(\frac{1}{a^2} + \frac{1}{b^2} + \frac{1}{c^2}\right) \geq \frac{9}{4}$$

We can rewrite the inequality as

$$S_a(b-c)^2 + S_b(c-a)^2 + S_c(a-b)^2 \geq 0$$

where

$$\begin{cases} S_a = \frac{2}{bc} - \frac{1}{a^2} \\ S_b = \frac{2}{ca} - \frac{1}{b^2} \\ S_c = \frac{2}{ab} - \frac{1}{c^2} \end{cases}$$

WLOG, suppose that $a \geq b \geq c$. Therefore, $S_a \geq 0$ and using **SOS theorem** *(fourth criterion)*, it is enough to check

$$b^2 S_b + c^2 S_c \geq 0$$

that can be reduced to

$$\frac{2(b^3 + c^3 - abc)}{abc} \geq 0$$

which is obvious because $a \leq b + c$ and as such

$$b^3 + c^3 \geq bc(b+c) \geq abc$$

Equality holds for $a = b = c$ or $a = b, c = 0$ or all permutations.

○ **Second proof:**

WLOG, we can assume that $x \geq y \geq z$. In this case, we get

$$4xy(x+y)^2 \geq 4y^2(x+z)^2$$
$$\geq (y+z)^2(x+z)^2$$

Therefore

$$\left(\frac{1}{x+z} - \frac{1}{y+z}\right)^2 = \frac{(x-y)^2}{(x+z)^2(y+z)^2}$$
$$\geq \frac{(x-y)^2}{4xy(x+y)^2}$$

which can be written as

$$\sum_{cyc} \frac{1}{(x+y)^2} \geq \frac{1}{4xy} + \frac{2}{(x+z)(y+z)}$$

To prove the original inequality, it is sufficient to prove

$$(xy + yz + zx)\left(\frac{1}{4xy} + \frac{2}{(x+z)(y+z)}\right) \geq \frac{9}{4}$$

$$\Longleftrightarrow \frac{1}{4} + \frac{z(x+y)}{4xy} + \frac{2(xy+yz+zx)}{(x+z)(y+z)} \geq \frac{9}{4}$$

$$\Longleftrightarrow \frac{z(x+y)}{4xy} + 2 - \frac{2z^2}{(x+z)(y+z)} \geq 2$$

$$\Longleftrightarrow \frac{z(x+y)}{4xy} \geq \frac{2z^2}{(x+z)(y+z)}$$

or equivalently

$$(x+y)(y+z)(z+x) \geq 8xyz$$

which is obviously true by **AM-GM inequality**. *Equality holds for $x = y = z$.*

○ *Third proof:*

The given inequality is equivalent to

$$4(xy + yz + zx)\left(\sum_{cyc}(x+y)^2(x+z)^2\right) \geq 9(x+y)^2(y+z)^2(z+x)^2$$

Let us denote $p = x + y + z$, $q = xy + yz + zx$ and $r = xyz$. We have

$$(x+y)^2(y+z)^2(z+x)^2 = (pq-r)^2$$

and

$$\sum_{cyc}(x+y)^2(x+z)^2 = (p^2+q)^2 - 4p(pq-r)$$

We can rewrite the original inequality as follows

$$4q((p^2+q)^2 - 4p(pq-r)) \geq 9(pq-r)^2$$

which is equivalent to

$$p^4q - 17p^2q^2 + 4q^3 + 34pqr - 9r^2 \geq 0$$

that can be transformed into

$$3pq(p^3 - 4pq + 9r) + q(p^4 - 5p^2q + 4q^2 + 6pr) + r(pq - 9r) \geq 0$$

*Now, using **Schur's inequality**, we deduce that*

$$p^3 - 4pq + 9r = \sum_{cyc} a(a-b)(a-c) \geq 0$$

$$p^4 - 5p^2q + 4q^2 + 6pr = \sum_{cyc} a^2(a-b)(a-b) \geq 0$$

*According to **AM-GM inequality***

$$pq = (a+b+c)(ab+bc+ca) \geq 9abc = 9r$$

Therefore

$$3pq(p^3 - 4pq + 9r) + q(p^4 - 5p^2q + 4q^2 + 6pr) + r(pq - 9r) \geq 0$$

as desired. Equality holds for $x = y = z$.

∎

Solution 33.

We will give two solutions to this inequality.

● *First Solution:*

We begin with the following identity

$$a^3 + b^3 + c^3 - 3abc = \frac{1}{2}(a+b+c)\left((a-b)^2 + (b-c)^2 + (c-a)^2\right)$$

Therefore

$$\frac{2\left(a^3 + b^3 + c^3\right) + 3abc}{a+b+c} = (a-b)^2 + (b-c)^2 + (c-a)^2 + \frac{9abc}{a+b+c}$$

It is easy to verify that

$$\frac{a^4 + b^4}{a^2 + b^2} - (a-b)^2 = \frac{ab\left(2a^2 + 2b^2 - 2ab\right)}{a^2 + b^2} \geq ab$$

Thus, we get

$$\sum_{cyc} \frac{a^4 + b^4}{a^2 + b^2} \geq \sum_{cyc} (a-b)^2 + (ab + bc + ca)$$

Now, we just need to check

$$ab + bc + ca \geq \frac{9abc}{a+b+c}$$

*Applying **AM-GM inequality**, we deduce*

$$(a+b+c)(ab+bc+ca) \geq 3\sqrt[3]{abc}.3\sqrt[3]{a^2b^2c^2} = 9abc$$

The inequality is proved. Equality holds for a = b = c.

• *Second Solution:*

The inequality is obviously true because of the following identity

$$LHS - RHS = \sum_{cyc} \frac{a(b^2 + c^2) + bc(b + c) + abc}{(b^2 + c^2)(a + b + c)}(b - c)^2$$

Equality at a = b = c.

∎

Solution 34.

*Applying **Complex Numbers** and the **Triangle inequality**, we deduce*

$$\sum_{cyc} \sqrt{1 + (a - b)^2} = |1 + i|a - b|| + |1 + i|b - c|| + |1 + i|c - a||$$

$$\geq |1 + i|a - b| + 1 + i|b - c| + 1 + i|c - a||$$
$$= |3 + i(|a - b| + |b - c| + |c - a|)|$$
$$= \sqrt{9 + (|a - b| + |b - c| + |c - a|)^2}$$
$$\geq \sqrt{9 + 2\left((a - b)^2 + (b - c)^2 + (c - a)^2\right)}$$

Hence, it is sufficient to prove

$$9 + 2\left((a - b)^2 + (b - c)^2 + (c - a)^2\right) \geq (a + b + c)^2$$
$$\Longleftrightarrow a^2 + b^2 + c^2 + 3 \geq 2(ab + bc + ca)$$

which is true because

$$a^2 + b^2 + c^2 + 3 = a^2 + b^2 + c^2 + \frac{3abc(a + b + c)}{ab + bc + ca}$$
$$\geq a^2 + b^2 + c^2 + \frac{9abc}{a + b + c}$$
$$\geq 2(ab + bc + ca)$$

*where the last transition is true by **Schur's inequality**.*

Comment 13.

*We will show in the following examples how **Complex Numbers** can be used to prove some Olympiad inequalities.*

○ *First example:*

Let P be an arbitrary point in the plane of triangle △ABC. Prove that:

$$a.PA^2 + b.PB^2 + c.PC^2 \geq abc$$

○ **Proof:**

Let us consider the origin of the complex plane at the point P and let x, y, z be the affixes of the vertices of the triangle $\triangle ABC$. The following identity is easy to verify

$$\sum_{cyc} \frac{x^2}{(x-y)(x-z)} = 1$$

By passing to moduli, it follows that

$$1 = \left| \sum_{cyc} \frac{x^2}{(x-y)(x-z)} \right| \leq \sum_{cyc} \left| \frac{x^2}{(x-y)(x-z)} \right|$$

Taking into account that $|x| = PA$, $|y| = PB$, $|z| = PC$ and $|y - z| = a$, $|x - z| = b$, $|x - y| = c$, the previous inequality is equivalent to

$$a.PA^2 + b.PB^2 + c.PC^2 \geq abc$$

as required.

○ **Second example:**

Let $a, b, c, d \geq 0$ such that $a^2 + b^2 + c^2 + d^2 = 1$. Prove that:

$$ab + bc + cd + da + ac + bd \leq \frac{5}{4} + 4abcd$$

○ **Proof:**

Let $S = ab + bc + cd + da + ac + bd$ and $P = abcd$. We need to prove

$$S \leq \frac{5}{4} + 4P$$

Consider the polynomial

$$f(x) = (x-a)(x-b)(x-c)(x-d)$$

We have

$$\begin{aligned}|f(it)|^2 &= |t^4 + i(a+b+c+d)t^3 - St^2 - i(abc + acd + abd + bcd)t + P| \\ &= (t^4 - St^2 + P)^2 + ((a+b+c+d)t^3 - (abc + acd + abd + bcd)t)^2 \\ &\geq (t^4 - St^2 + P)^2\end{aligned}$$

On the other hand, applying **AM-GM inequality**

$$\begin{aligned}|f(it)|^2 &= (t^2 + a^2)(t^2 + b^2)(t^2 + c^2)(t^2 + d^2) \\ &\leq \frac{1}{256}\left(4t^2 + a^2 + b^2 + c^2 + d^2\right)^4 \\ &= \frac{1}{256}(4t^2 + 1)^4\end{aligned}$$

Therefore

$$(t^4 - St^2 + P)^2 \leq \frac{1}{256}(4t^2 + 1)^4$$

For $t = \frac{1}{2}$, we get

$$\left|\frac{1}{16} - \frac{S}{4} + P\right| \leq \frac{1}{4}$$

or equivalently

$$\left|\frac{1}{4} - S + 4P\right| \leq 1$$

Consequently

$$S \leq \frac{5}{4} + 4P$$

as desired. Equality holds for $a = b = c = d = \frac{1}{2}$.

∎

Solution 35.

We need to prove that

$$3\sqrt{3(a^2 + b^2 + c^2)} \geq \sqrt[3]{9(a^3 + b^3 + c^3)} + 2(a + b + c)$$

Or

$$3\left(\sqrt{3(a^2 + b^2 + c^2)} - a - b - c\right) \geq \sqrt[3]{9(a^3 + b^3 + c^3)} - a - b - c$$

which is equivalent to

$$\frac{6\sum_{cyc}(a^2 - ab)}{\sqrt{3(a^2 + b^2 + c^2)} + a + b + c} \geq \frac{\sum_{cyc}(8a^3 - 3a^2b - 3a^2c - 2abc)}{9\left(\left(\sqrt[3]{\frac{a^3+b^3+c^3}{3}}\right)^2 + \sqrt[3]{\frac{a^3+b^3+c^3}{3}}\frac{a+b+c}{3} + \left(\frac{a+b+c}{3}\right)^2\right)}$$

Now, let $\sqrt{\frac{a^2+b^2+c^2}{3}} = t\frac{a+b+c}{3}$. Therefore, $t \geq 1$ and since

$$\sqrt[3]{\frac{a^3 + b^3 + c^3}{3}} \geq \sqrt{\frac{a^2 + b^2 + c^2}{3}}$$

$$\sum_{cyc}(8a^3 - 3a^2b - 3a^2c - 2abc) \leq 8\sum_{cyc}(a^3 - abc)$$

It is enough to prove that

$$\frac{6}{3(t+1)} \geq \frac{24}{9(t^2 + t + 1)}$$

which is equivalent to

$$3t^2 - t - 1 \geq 0$$

This inequality is obviously true. Equality holds for $a = b = c$.

∎

Solution 36.

We will give two solutions to this problem.

● *First Solution:*

The inequality is homogeneous, we can therefore assume $a + b + c = 1$.

On the other hand, we know that

$$\frac{a^3}{(b+c)^2} + b + c = \frac{a^3 + b^3 + c^3}{(b+c)^2} + \frac{3bc}{b+c}$$

After adding $2(a+b+c) = 2$ to both sides, we can rewrite the inequality as

$$\left(\sum_{cyc} a^3\right)\left(\sum_{cyc} \frac{1}{(a+b)^2}\right) + 3\left(\sum_{cyc} \frac{ab}{a+b}\right) \geq \frac{11\left(\sum_{cyc} a^2\right) - 8\left(\sum_{cyc} ab\right)}{4(a+b+c)} + 2$$

Let $p = a+b+c$, $q = ab+bc+ca$ and $r = abc$, we get

$$\frac{(p^3 - 3pq + 3r)\left[(p^2+q)^2 - 4p(pq-r)\right]}{(pq-r)^2} + \frac{3(q^2+pr)}{pq-r} \geq \frac{11p^2 - 30q}{4p} + 2$$

On the other hand $p = 1$, thus we need to prove

$$f(r) = (1 - 3q + 3r)\left(\frac{(1-q)^2 + 4r}{(q-r)^2}\right) + 3\left(\frac{q^2+r}{q-r}\right) \geq \frac{19}{4} - \frac{15q}{2}$$

f is obviously an increasing function on $\mathbb{R}^+$ and we need to consider 2 cases:

(i) **The first case:** $0 < q \leq \frac{4\sqrt{6}-6}{15}$.

    In this particular case, we get

$$f(r) \geq f(0)$$
$$= \frac{(1-3q)(1-q)^2}{q^2} + 3q$$
$$= \frac{19}{4} - \frac{15q}{2} + \frac{(2q-1)(15q^2 + 12q - 4)}{4q^2}$$
$$\geq \frac{19}{4} - \frac{15q}{2}$$

(ii) **The second case:** $\frac{4\sqrt{6}-6}{15} \leq q \leq \frac{1}{3}$.

    Applying **Schur's inequality**, we get

$$r \geq \frac{(4q - p^2)(p^2 - q)}{6} = \frac{(4q-1)(1-q)}{6}$$

Therefore

$$f(r) \geq f\left(\frac{(4q-1)(1-q)}{6}\right)$$

$$= \frac{6(1-q-4q^2)(1-q)(5q+1)}{(4q^2+q+1)^2} + 3\left(\frac{2q^2+5q-1}{4q^2+q+1}\right)$$

It remains to prove that

$$\frac{6(1-q-4q^2)(1-q)(5q+1)}{(4q^2+q+1)^2} + 3\left(\frac{2q^2+5q-1}{4q^2+q+1}\right) \geq \frac{19}{4} - \frac{15q}{2}$$

After simplifying the previous expression, we get

$$\frac{(3q-1)\left(160q^4 + 224q^3 + 114q^2 - 91q + 7\right)}{(4q^2+q+1)^2} \geq 0$$

which is true for $\frac{4\sqrt{6}-6}{15} \leq q \leq \frac{1}{3}$.

The proof is complete. Equality holds for $a = b = c$.

- *Second Solution:*

WLOG, we can assume that $a + b + c = 3$. Let

$$L(a,b,c) = \frac{a^3}{(b+c)^2} + \frac{b^3}{(c+a)^2} + \frac{c^3}{(a+b)^2} - \frac{11\left(a^2+b^2+c^2\right) - 8\left(ab+bc+ca\right)}{4(a+b+c)}$$

We have

$$L(a,b,c) = \sum_{cyc}\left(\frac{a^3}{(3-a)^2} - \frac{11a^2 - 4(ab+ac)}{12}\right)$$

$$= \sum_{cyc}\left(\frac{a^3}{(3-a)^2} - \frac{11a^2 - 4a(3-a)}{12}\right)$$

$$= \sum_{cyc}\left(\frac{a^3}{(3-a)^2} + a - \frac{5a^2}{4}\right)$$

Let

$$f(u) = \frac{u^3}{(3-u)^2} + u - \frac{5u^2}{4}$$

We can rewrite the inequality as follows

$$f(a) + f(b) + f(c) \geq 3f(1)$$

On the other hand, we have

$$f''(u) = \frac{54u}{(3-u)^4} - \frac{5}{2}$$

*f'' admits a unique root x_0 on $[0,3[$ with $x_0 < 1$. Therefore, f is convex on $[1,3[$ and according to **Vasc's HCF theorem**, it is enough to check the case*

$$f(y) + 2f(x) \geq 3f(1)$$

with $y + 2x = 3$.

After some simplifications, we get

$$\frac{3(x-1)^2(81 - 54x - 63x^2 + 60x^3 - 10x^4)}{4(3-x)^2 x^2} \geq 0$$

which is obviously true. Equality holds for $a = b = c$.

∎

Solution 37.

*Applying **Cauchy-Schwarz inequality**, we deduce that*

$$(a^3 + b^3 + c^3)(a + b + c) \geq (a^2 + b^2 + c^2)^2$$

*Applying **Weighted AM-GM inequality**, we obtain that*

$$\sqrt[3]{\frac{a^3 + b^3 + c^3}{3}} + \frac{a+b+c}{9} = \frac{4}{3}\left(\frac{3}{4}\sqrt[3]{\frac{a^3+b^3+c^3}{3}} + \frac{1}{4}\left(\frac{a+b+c}{3}\right)\right)$$

$$\geq \frac{4}{3}\left(\frac{a^3+b^3+c^3}{3}\right)^{\frac{1}{4}}\left(\frac{a+b+c}{3}\right)^{\frac{1}{4}}$$

$$\geq \frac{4}{3}\left(\frac{a^2+b^2+c^2}{3}\right)^{\frac{1}{2}}$$

$$= \frac{4}{3}$$

as required. Equality holds for $a = b = c = 1$.

∎

Solution 38.

We will prove the following stronger inequality

$$\frac{1}{a^2 + b^2} + \frac{1}{b^2 + c^2} + \frac{1}{c^2 + a^2} \geq \frac{5 + abc}{4}$$

which is equivalent to prove

$$(a^2 + b^2 + c^2 + abc)\left(\frac{1}{a^2 + b^2} + \frac{1}{b^2 + c^2} + \frac{1}{c^2 + a^2}\right) \geq 5 + abc$$

$$\Longleftrightarrow \sum_{cyc} \frac{c^2}{a^2 + b^2} + abc\left(\sum_{cyc} \frac{1}{a^2 + b^2}\right) - abc \geq 2$$

*According to **Schur's inequality**, we obtain*

$$\sum_{cyc} \frac{c^2}{a^2 + b^2} - 2 = \frac{a^6 + b^6 + c^6 - a^2 b^2 c^2 - \sum_{cyc} a^2 b^2 (a^2 + b^2)}{(a^2 + b^2)(b^2 + c^2)(c^2 + a^2)}$$

$$\geq \frac{-4a^2b^2c^2}{(a^2+b^2)(b^2+c^2)(c^2+a^2)}$$

Therefore, we only need to prove

$$\sum_{cyc} \frac{1}{a^2+b^2} - 1 \geq \frac{4abc}{(a^2+b^2)(b^2+c^2)(c^2+a^2)}$$

which is equivalent to

$$(4-abc)^2 + (abc-3)\left(\sum_{cyc} a^2b^2\right) + a^2b^2c^2 - 4abc \geq 0$$

On the other hand, we have

$$abc \leq 1$$

$$\sum_{cyc} a^2b^2 \leq \frac{\left(\sum_{cyc} a^2\right)^2}{3} = \frac{(4-abc)^2}{3}$$

Consequently, we only need to check

$$\frac{abc(1-abc)(4-abc)}{3} \geq 0$$

which is obviously true.

■

Solution 39.

First of all, we will prove the following classic inequality

$$\frac{a^2}{b^2+c^2} + \frac{b^2}{c^2+a^2} + \frac{c^2}{a^2+b^2} \geq \frac{a}{b+c} + \frac{b}{c+a} + \frac{c}{a+b}$$

In fact, we have

$$\sum_{cyc} \left(\frac{a^2}{b^2+c^2} - \frac{a}{b+c} \right) = \sum_{cyc} \frac{ab(a-b)+ac(a-c)}{(b+c)(b^2+c^2)}$$

$$= \sum_{cyc} \left[\frac{ab(a-b)}{(b+c)^2(b^2+c^2)} - \frac{ab(a-b)}{(c+a)(c^2+a^2)} \right]$$

$$= \sum_{cyc} \frac{ab(a-b)^2(a^2+b^2+c^2+ab+bc+ca)}{(b+c)(c+a)(b^2+c^2)(c^2+a^2)}$$

$$\geq 0$$

The inequality is proved.

Let $m = \frac{a^2+b^2+c^2}{ab+bc+ca}$. *Using the above-mentioned inequality, we get*

$$\sum_{cyc} \frac{2(a^2+bc)}{b^2+c^2} = \sum_{cyc} \frac{2a^2}{b^2+c^2} + \sum_{cyc} \frac{2bc}{b^2+c^2}$$

$$\geq \sum_{cyc} \frac{2a}{b+c} + \sum_{cyc} \frac{2bc}{b^2+c^2}$$

$$= \sum_{cyc} \frac{2a}{b+c} + \sum_{cyc} \frac{(b+c)^2}{b^2+c^2} - 3$$

$$\geq \frac{(a+b+c)^2}{ab+bc+ca} + \frac{2(a+b+c)^2}{a^2+b^2+c^2} - 3$$

$$= \frac{a^2+b^2+c^2}{ab+bc+ca} + \frac{4(ab+bc+ca)}{a^2+b^2+c^2} + 1$$

$$= m + \frac{4}{m} + 1$$

$$= \frac{(m-2)^2}{m} + 5$$

$$\geq 5$$

as required. Equality holds for $a = b, c = 0$ and permutations.

Comment 14.

We can also use the **Derivatives method** to prove

$$\frac{a^2}{b^2+c^2} + \frac{b^2}{c^2+a^2} + \frac{c^2}{a^2+b^2} \geq \frac{a}{b+c} + \frac{b}{c+a} + \frac{c}{a+b}$$

In fact, it is enough to prove that the following function is increasing

$$f(x) = \frac{a^x}{b^x+c^x} + \frac{b^x}{c^x+a^x} + \frac{c^x}{a^x+b^x}$$

Calculating the derivative, we get

$$f'(x) = \sum_{cyc} \frac{a^x \ln a(b^x+c^x) - a^x(b^x \ln b + c^x \ln c)}{(b^x+c^x)^2}$$

$$= \sum_{cyc} \frac{a^x(b^x(\ln a - \ln b) - c^x(\ln c - \ln a))}{(b^x+c^x)^2}$$

$$= \sum_{cyc} (\ln a - \ln b) \left(\frac{a^x b^x}{(b^x+c^x)^2} - \frac{a^x b^x}{(c^x+a^x)^2} \right)$$

$$= \sum_{cyc} \frac{a^x b^x (a^x + b^x + 2c^x)(a^x - b^x)(\ln a - \ln b)}{(b^x+c^x)^2(c^x+a^x)^2}$$

$$\geq 0$$

Consequently, f is an increasing function. Therefore

$$f(2) \geq f(1)$$

which is exactly

$$\frac{a^2}{b^2+c^2} + \frac{b^2}{c^2+a^2} + \frac{c^2}{a^2+b^2} \geq \frac{a}{b+c} + \frac{b}{c+a} + \frac{c}{a+b}$$

Equality holds for $a = b = c$.

Solution 40.

Notice

$$LHS = \sum_{cyc} \frac{a^3}{a+b} + abc\left(\sum_{cyc} \frac{1}{a+b}\right) \geq \frac{(a^2+b^2+c^2)^2}{a^2+b^2+c^2+ab+bc+ca} + \frac{9abc}{2(a+b+c)}$$

Let s and t be non-negative real numbers such that $a+b+c = 3s$ and $ab+bc+ca = 3(s^2-t^2)$. There are 2 cases to consider:

(i) **The first case:** $2t \geq s$.

It is enough to prove

$$\frac{(a^2+b^2+c^2)^2}{a^2+b^2+c^2+ab+bc+ca} \geq \frac{(a+b+c)^2}{3}$$

On the other hand, we have

$$\frac{(a^2+b^2+c^2)^2}{a^2+b^2+c^2+ab+bc+ca} - \frac{(a+b+c)^2}{3} = \frac{3(s^2+2t^2)^2}{2s^2+t^2} - 3s^2$$

$$= \frac{3(s^2+t^2)(2t-s)(s+2t)}{2s^2+t^2}$$

$$\geq 0$$

Therefore

$$\frac{(a^2+b^2+c^2)^2}{a^2+b^2+c^2+ab+bc+ca} + \frac{9abc}{2(a+b+c)} \geq \frac{(a+b+c)^2}{3}$$

(ii) **The second case:** $s \geq 2t$.

Using **Theorem 39**, *we get*

$$abc \geq (s-2t)(s+t)^2$$

Consequently, we only need to prove

$$\frac{3(s^2+2t^2)^2}{2s^2+t^2} + \frac{3(s-2t)(s+t)^2}{2s} \geq 3s^2$$

After simplifications, we get

$$\frac{3t^2(s-2t)(s-t)^2}{2s(2s^2+t^2)} \geq 0$$

which is obviously true.

The proof is complete. Equality holds for $a = b = c = 1$.

Comment 15.

*We will show here a very elegant and short proof of **Theorem 39**.*

Let $p = a + b + c$, $q = ab + bc + ca$ and $r = abc$. We have

$$(a-b)^2(b-c)^2(c-a)^2 = -4p^3r + p^2q^2 + 18pqr - 4q^3 - 27r^2$$

Using $a + b + c = 3s$ and $ab + bc + ca = 3(s^2 - t^2)$, we get

$$(a-b)^2(b-c)^2(c-a)^2 = -4p^3r + p^2q^2 + 18pqr - 4q^3 - 27r^2$$
$$= -27(r - (s-t)^2(s+2t))(r - (s-2t)(s+t)^2)$$
$$\geq 0$$

Therefore, we conclude that

$$(s - 2t)(s + t)^2 \leq r \leq (s + 2t)(s - t)^2$$

■

Solution 41.

We will give two solutions to this problem.

- *First Solution:*

*We will use the **Mixing Variables method**. Define*

$$N = 5\left(a^2 + 6ab + b^2\right) - 48$$

$$Q = 5\left(b^2 + 6bc + c^2\right) - 48$$

$$T = 5\left(c^2 + 6ca + a^2\right) - 48$$

On the other hand, we have

$$N + Q + T = 10\left[(a + b + c)^2 + ab + bc + ca\right] - 144$$
$$\leq 10\,(9 + 3) - 144$$
$$< 0$$

Consequently, at least one of the three expressions N, Q, T is negative. WLOG, let's assume that $Q < 0$.

Let

$$f(a, b, c) = (5a^2 + 6)(5b^2 + 6)(5c^2 + 6)$$

Therefore

$$f(a, b, c) - f\left(a, \frac{b+c}{2}, \frac{b+c}{2}\right) = -\frac{5}{16} \cdot (5a^2 + 6) \cdot Q \cdot (b - c)^2 \geq 0$$

Let $t = \frac{b+c}{2} = \frac{3-a}{2}$. We only need to prove

$$f(a, t, t) \geq 1331$$

which is equivalent to

$$(a-1)^2 \left(25a^4 - 250a^3 + 1095a^2 - 2060a + 1454\right) \geq 0$$

This last inequality is obviously true. Equality holds for $a = b = c = 1$.

• *Second Solution:*

Let $f(x) = \ln(1 + x^2)$ for $x \in [0, 3]$. The inequality is equivalent to prove

$$f(a) + f(b) + f(c) \geq 3f(1)$$

We have

$$f''(x) = \frac{60 - 50x^2}{(5x^2 + 6)^2}$$

Therefore, f is convex on $[0, 1]$. According to **Vasc's HCF theorem**, it is sufficient to prove

$$f(y) + 2f(x) \geq 3f(1)$$

with $y + 2x = 3$.

After some simplifications, we need to prove

$$5(x-1)^2(100x^4 - 100x^3 + 195x^2 - 230x + 101) \geq 0$$

which is true. Equality holds for $a = b = c = 1$.

∎

Solution 42.

From the well-known inequality

$$(x + y + z)^2 \geq 3(xy + yz + zx)$$

we get

$$\sum_{cyc} a^2 b^2 \geq \sqrt{3a^2 b^2 c^2 (a^2 + b^2 + c^2)} = 3abc$$

Using this inequality and **Schur's inequality**, we get

$$\sum_{cyc} a^4 b^4 + 3(abc)^3 = \sum_{cyc} a^4 b^4 + \frac{9(abc)^4}{3abc}$$

$$\geq \sum_{cyc} a^4 b^4 + \frac{9(abc)^4}{\sum a^2 b^2}$$

$$\geq \sum_{cyc} 2(ab)^2 (bc)^2$$

$$= 6a^2 b^2 c^2$$

This ends the proof. Equality holds for $(1, 1, 1)$, $(\sqrt{3}, 0, 0)$ and permutations.

Solution 43.

Let $a^3 + b^3 + c^3 = 3u$, $a^3b^3 + b^3c^3 + c^3a^3 = 3v^2$ and $a^3b^3c^3 = w^3$. Thus, the condition gives

$$u + w = 2$$

We need to prove

$$\frac{v^2}{w^3} + 1 \geq 2u$$

Or

$$v^2(u+w)^2 \geq 2w^3(3u - w)$$

Or

$$u^2v^2 + 2uv^2w + v^2w^2 + 2w^4 \geq 6uw^3$$

which can easily be proved by **AM-GM inequality**

$$u^2v^2 + 2uv^2w + v^2w^2 + 2w^4 \geq 6\sqrt[6]{u^2v^2\left(uv^2w\right)^2 v^2w^2 \left(w^4\right)^2}$$
$$= 6\sqrt[6]{u^4v^8w^{12}}$$
$$\geq 6\sqrt[6]{u^4\left(uw^3\right)^2 w^{12}}$$
$$= 6uw^3$$

as required. Equality holds for $a = b = c = 1$.

■

Solution 44.

○ **Right inequality:**

Using **Cauchy-Schwarz inequality**, we find that

$$\frac{a}{b+c} + \frac{b}{c+a} + \frac{c}{a+b} \geq \frac{(a+b+c)^2}{2(ab+bc+ca)} = \frac{a+b+c}{2}$$

○ **Left inequality:**

By adding 3 to both sides, we need to prove

$$a + b + c + \frac{3}{2} \geq (a+b+c)\left(\frac{1}{b+c} + \frac{1}{c+a} + \frac{1}{a+b}\right)$$

Let $t = a + b + c \geq 3$. We have

$$(a+b+c)\left(\frac{1}{b+c} + \frac{1}{c+a} + \frac{1}{a+b}\right) = \frac{t(t^2+t)}{t^2 - abc}$$
$$\leq \frac{t(t^2+t)}{t^2 - \frac{(ab+bc+ca)^2}{3(a+b+c)}}$$
$$= \frac{t(t^2+t)}{t^2 - \frac{t}{3}}$$

$$= t + \frac{4t}{3t - 1}$$

$$= t + \frac{3}{2} - \frac{t - 3}{2(3t - 1)}$$

$$\leq t + \frac{3}{2}$$

$$= a + b + c + \frac{3}{2}$$

The proof is complete. Equality holds for a = b = c = 1.

∎

Solution 45.

We will give two solutions to this inequality.

- ***First Solution:***

We can rewrite the initial condition as

$$\frac{a}{a + 2} + \frac{b}{b + 2} + \frac{c}{c + 2} = 1$$

*Applying **Cauchy-Schwarz inequality**, we find that*

$$1 = \frac{a}{a + 2} + \frac{b}{b + 2} + \frac{c}{c + 2} \geq \frac{(\sqrt{a} + \sqrt{b} + \sqrt{c})^2}{a + b + c + 6}$$

which is equivalent to

$$\sqrt{ab} + \sqrt{bc} + \sqrt{ca} \leq 3$$

The proof is complete. Equality holds for a = b = c = 1.

- ***Second Solution:***

We have ab + bc + ca + abc = 4, therefore, there exist x, y, z such that

$$\begin{cases} a = \frac{2x}{y+z} \\ b = \frac{2y}{z+x} \\ c = \frac{2z}{x+y} \end{cases}$$

The inequality becomes

$$2 \sum_{cyc} \sqrt{\frac{xy}{(y + z)(z + x)}} \leq 3$$

*which is true by **AM-GM inequality***

$$2 \sum_{cyc} \sqrt{\frac{xy}{(y + z)(z + x)}} \leq \sum_{cyc} \left(\frac{x}{z + x} + \frac{y}{y + z} \right) = 3$$

as desired. Equality holds for a = b = c = 1.

Solution 46.

We will prove this inequality in two steps. First, we will prove that

$$\frac{2}{ab+bc+ca} + \frac{1}{a^2+b^2+c^2} \geq \frac{a^2+b^2+c^2+3}{6}$$

then

$$\frac{1}{a^2+2bc} + \frac{1}{b^2+2ca} + \frac{1}{c^2+2ab} \geq \frac{2}{ab+bc+ca} + \frac{1}{a^2+b^2+c^2}$$

*Let $m = a^2 + b^2 + c^2$, $n = ab + bc + ca$ and $p = a + b + c$. According to **Problem 2** and **Problem 4**, we have*

$$m \geq 3 \geq p$$

On the other hand, we have

$$2mn = 2(a^2+b^2+c^2)(ab+bc+ca)$$

$$= 2\sum_{cyc} a^3(b+c) + 2abc\left(\sum_{cyc} a\right)$$

$$\leq \sum_{cyc} a^4 + \sum_{cyc} a^3(b+c) + 3abc\left(\sum_{cyc} a\right)$$

$$= 6p$$

Now, our inequality

$$\frac{2}{ab+bc+ca} + \frac{1}{a^2+b^2+c^2} \geq \frac{a^2+b^2+c^2+3}{6}$$

is equivalent to

$$6(2m+n) \geq mn(m+3)$$

On the other hand, we proved that $mn \leq 3p$. Consequently, it is sufficient to prove

$$2(2m+n) \geq p(m+3)$$

Or

$$3m + p^2 \geq p(m+3)$$

Or

$$(m-p)(3-p) \geq 0$$

which is obviously true.

Now, let's prove

$$\frac{1}{a^2+2bc} + \frac{1}{b^2+2ca} + \frac{1}{c^2+2ab} \geq \frac{2}{ab+bc+ca} + \frac{1}{a^2+b^2+c^2}$$

We have the following identity

$$\sum_{cyc}(a^2+2bc)(b^2+2ca) = (ab+bc+ca)(2(a^2+b^2+c^2)+ab+bc+ca)$$

Consequently, we can rewrite the original inequality as

$$\frac{ab + bc + ca}{(a^2 + 2bc)(b^2 + 2ca)(c^2 + 2ab)} \geq \frac{1}{(a^2 + b^2 + c^2)(ab + bc + ca)}$$

which is true because of the following identity

$$(a^2 + b^2 + c^2)(ab + bc + ca)^2 - (a^2 + 2bc)(b^2 + 2ca)(c^2 + 2ab) = (a - b)^2(b - c)^2(c - a)^2$$

Finally, the original inequality is true with equality at $a = b = c = 1$.

■

Solution 47.

Using **Derivatives**, *we can easily prove that for any real number x, we have*

$$e^x \geq 1 + x + \frac{x^2}{2} + \frac{x^3}{6}$$

Let $t = \ln 2$. Therefore, by taking $x = t(a - 1)$, we get

$$2^{a+1} \geq 4 + 4t(a - 1) + 2t^2(a - 1)^2 + \frac{2}{3}t^3(a - 1)^3$$

Thus, it is enough to prove that

$$\sum_{cyc}\left(4 + 4t(a - 1) + 2t^2(a - 1)^2 + \frac{2}{3}t^3(a - 1)^3\right) + abc \geq 13$$

which is a linear inequality of w^3. By **uvw method**, *it is enough to prove the last inequality in the following cases:*

(i) **The first case:** $b = a$ *and $c = 3 - 2a$, where $0 < a < \frac{3}{2}$.*

We need to prove that

$$2t^2(2(a - 1)^2 + (3 - 2a - 1)^2) + \frac{2}{3}t^3(2(a - 1)^3 + (3 - 2a - 1)^3) + a^2(3 - 2a) - 1 \geq 0$$

Or

$$(a - 1)^2(-4at^3 - 2a + 4t^3 + 12t^2 - 1)$$

Or

$$a \leq \frac{12t^2 + 4t^3 - 1}{2(2t^3 + 1)} \approx 1.829...$$

which is true.

(ii) **The second case:** $w^3 \to 0^+$.

Let $c \to 0^+$ and $b = 3 - a$, where $0 < a < 3$. We need to prove that

$$2t^2(1 + (a - 1)^2 + (2 - a)^2) + \frac{2}{3}t^3((a - 1)^3 + (2 - a)^3 - 1) - 1 \geq 0$$

Or equivalently

$$t^2(t + 2)(2x - 3)^2 - t^3 + 6t^2 - 2 \geq 0$$

which is true as $-t^3 + 6t^2 - 2 \approx 0.2748... > 0$. The proof is now complete.

Comment 16.

If you are wondering how we came up with the expression $1 + x + \frac{x^2}{2} + \frac{x^3}{6}$, *the answer is pretty simple:* **Taylor Expansion formula** *which is a one-dimensional expansion of a real function f as follows*

$$f(x) = \sum_{k=0}^{+\infty} \frac{f^{(k)}(0)}{k!} x^k$$

In the particular case of $f(x) = e^x$, *we get*

$$e^x = \sum_{k=0}^{+\infty} \frac{x^k}{k!}$$

We will show here how to prove

$$e^x \geq 1 + x + \frac{x^2}{2} + \frac{x^3}{6}$$

○ **First proof:**

Let

$$f(x) = e^x - 1 - x - \frac{x^2}{2} - \frac{x^3}{6}$$

It suffices to show that $f(x) \geq 0$ *for all x.*

First, we have

$$\begin{cases} f'(x) = e^x - 1 - x - \frac{x^2}{2} \\ f''(x) = e^x - 1 - x \\ f'''(x) = e^x - 1 \end{cases}.$$

We see that $f'''(x) = e^x - 1$ *is a strictly increasing function, whose only zero is at* $x = 0$. *Therefore,* $f''(x)$ *is a strictly convex function whose minimum is at* $x = 0$. *This shows that* $f''(x) \geq 0$ *for all x, and equality occurs if and only if* $x = 0$.

From this, we can say that $f'(x)$ *is an increasing function, which means that it can have at most one zero. This zero is at* $x = 0$.

Finally, since $f'(x)$ *is an increasing function whose only zero is at* $x = 0$, *this means that* $f(x)$ *is a strictly convex function whose minimum is at* $x = 0$.

This shows that

$$e^x \geq 1 + x + \frac{x^2}{2} + \frac{x^3}{6}$$

Equality occurs if and only if $x = 0$.

○ **Second proof:**

Consider the function

$$f(x) = e^{-x} \left(1 + x + \frac{x^2}{2} + \frac{x^3}{6} \right)$$

It suffices to show that $f(x) \leq 1$, i.e. the maximum of $f(x)$ is 1.

Checking the boundaries: $\lim_{x \to +\infty} f(x) = 0$ *and* $\lim_{x \to -\infty} f(x) = -\infty$.

By taking derivatives, we get

$$f'(x) = -\frac{1}{6}x^3 e^{-x}$$

which is 0 if and only if $x = 0$.

Now, it is clear the maximum value occurs at $f(0) = 1$.

○ ***Third proof:***

Let $f(x) = e^x$. *According to **Taylor's Theorem with Lagrange Remainder**, for all $x \in \mathbb{R}$, there exists θ_x such that*

$$e^x = 1 + x + \frac{x^2}{2} + \frac{x^3}{6} + \frac{e^{\theta_x} x^4}{24}$$

Therefore

$$e^x \geq 1 + x + \frac{x^2}{2} + \frac{x^3}{6}$$

as required.

■

Solution 48.

WLOG, we can assume that $a \geq b \geq c$ and we have

$$\prod_{cyc}(a+b)\sum_{cyc}c(a-b)^2 - 12\sum_{cyc}(a^2-ab) = \sum_{cyc}\left(a\prod_{cyc}(a+b) - 6\right)(b-c)^2$$

Let

$$\begin{cases} S_a = \left(a\prod_{cyc}(a+b) - 6\right) \\ S_b = \left(b\prod_{cyc}(a+b) - 6\right) \\ S_c = \left(c\prod_{cyc}(a+b) - 6\right) \end{cases}$$

It is sufficient to prove

$$\sum_{cyc} S_a(b-c)^2 \geq 0$$

On the other hand, we know that $a \geq b \geq c$, therefore

$$S_a \geq S_b \geq S_c$$

Furthermore

$$\begin{aligned} S_b + S_c &= (a+b)(a+c)(b+c)^2 - 12 \\ &= (a^2 + ab + bc + ca)(b+c)^2 - 12 \end{aligned}$$

$$\geq a(a+b+c)(b+c)^2 - 12$$
$$\geq 4abc(a+b+c) - 12$$
$$= 0$$

Applying **SOS theorem**, we conclude that

$$\sum_{cyc} S_a(b-c)^2 \geq 0$$

as desired. Equality holds for $a = b = c = 1$.

■

Solution 49.

Using the **Buffalo Way technique**: let $a = \min\{a, b, c\}$, $b = a + u$, $c = a + v$ and $u^2 + v^2 = 2kuv$. Thus, $k > 1$ and we have

$$(a+b+c)\sum_{cyc}\frac{a+b}{(a-b)^2} = (3a+u+v)\left(\frac{2a+u}{u^2} + \frac{2a+v}{v^2} + \frac{2a+u+v}{(u-v)^2}\right)$$

$$\geq (u+v)\left(\frac{1}{u} + \frac{1}{v} + \frac{u+v}{(u-v)^2}\right)$$

$$= 2k + 2 + \frac{2k+2}{2k-2}$$

$$= 9 + \frac{2(k-2)^2}{k-1}$$

$$\geq 9$$

as required.

■

Solution 50.

The inequality can be transformed into

$$6 + (ab)^3 + (bc)^3 + (ca)^3 + a^3 + b^3 + c^3 \geq 4(\sqrt{a^3} + \sqrt{b^3} + \sqrt{c^3})$$

Applying **Schur's inequality**, we obtain

$$3 + (ab)^3 + (bc)^3 + (ca)^3 \geq \sum_{cyc} ab^2c(ab+bc)$$

$$= \sum_{cyc} a^2(b+c)$$

Therefore

$$LHS \geq 3 + (a^2 + b^2 + c^2)(a+b+c)$$

According to **Cauchy-Schwarz inequality**, we find that

$$\sqrt{(a^2 + b^2 + c^2)(a+b+c)} \geq \sqrt{a^3} + \sqrt{b^3} + \sqrt{c^3}$$

Consequently, it is sufficient to prove

$$3 + (a^2 + b^2 + c^2)(a + b + c) \geq 4\sqrt{(a^2 + b^2 + c^2)(a + b + c)}$$

which is equivalent to

$$\left(\sqrt{(a^2 + b^2 + c^2)(a + b + c)} - 1\right)\left(\sqrt{(a^2 + b^2 + c^2)(a + b + c)} - 3\right) \geq 0$$

*This last inequality is obviously true by **AM-GM inequality***

$$(a^2 + b^2 + c^2)(a + b + c) \geq 3\sqrt[3]{a^2 b^2 c^2} \cdot 3\sqrt[3]{abc}$$
$$= 9abc$$
$$= 9$$

as desired. Equality holds for $a = b = c = 1$.

∎

Solution 51.

We will give three solutions to this inequality.

● ***First Solution:***

We should prove that

$$2(a + b + c) + 2\left(\sum_{cyc}\sqrt{(a + b)(b + c)}\right) \geq 18$$

On the other hand, we know that

$$a + b + c \geq \sqrt{3(ab + bc + ca)} = 3$$

$$\sqrt{(a + b)(b + c)} = \sqrt{b^2 + 3}$$

Consequently, it is sufficient to prove

$$\sum_{cyc}\sqrt{a^2 + 3} \geq 6$$

*which is true by **Minkowski's inequality***

$$\sum_{cyc}\sqrt{a^2 + 3} \geq \sqrt{(a + b + c)^2 + (3\sqrt{3})^2} \geq 6$$

The proof is complete. Equality holds for $a = b = c = 1$.

● ***Second Solution:***

*Applying **AM-GM inequality**, we obtain*

$$\sqrt{a + b} + \sqrt{b + c} + \sqrt{c + a} \geq 3\sqrt[6]{(a + b)(b + c)(c + a)}$$

On the other hand, we know that

$$(a+b)(b+c)(c+a) \geq \frac{8}{9}(a+b+c)(ab+bc+ca)$$

$$a+b+c \geq \sqrt{3(ab+bc+ca)}$$

Finally

$$\sqrt{a+b} + \sqrt{b+c} + \sqrt{c+a} \geq 3\sqrt[6]{\frac{8}{3}\cdot\sqrt{3(ab+bc+ca)}} = 3\sqrt{2}$$

as desired. Equality holds for $a=b=c=1$.

- ***Third Solution:***

Let's consider the triangle with sides $(\sqrt{a+b}, \sqrt{b+c}, \sqrt{c+a})$ and area S.

*From **Heron's formula**, we get*

$$S = \frac{1}{2}\sqrt{ab+bc+ca}$$

*Applying **Hadwiger-Finsler's inequality**, we deduce that*

$$\sum_{cyc}(\sqrt{a+b})^2 \geq \sum_{cyc}(\sqrt{a+b}-\sqrt{a+c})^2 + 4\sqrt{3}S$$

Therefore, we get the well-known inequality

$$\sum_{cyc}\sqrt{(a+b)(a+c)} \geq a+b+c + \sqrt{3}\cdot\sqrt{ab+bc+ca}$$

Using $(x+y+z)^2 \geq 3(xy+yz+zx)$ and the above-mentioned inequality, we deduce that

$$\sqrt{a+b}+\sqrt{b+c}+\sqrt{c+a} \geq \sqrt{3\sum_{cyc}\sqrt{a+b}\sqrt{a+c}}$$
$$\geq \sqrt{3(a+b+c+\sqrt{3}\cdot\sqrt{ab+bc+ca})}$$
$$\geq \sqrt{3(\sqrt{3}\cdot\sqrt{ab+bc+ca}+\sqrt{3}\cdot\sqrt{ab+bc+ca})}$$
$$= \sqrt{6\sqrt{3}\cdot\sqrt{ab+bc+ca}}$$
$$= 3\sqrt{2}$$

The proof is complete. Equality holds for $a=b=c=1$.

Comment 17.

*We will present here a nice proof of **Hadwiger-Finsler's inequality** using **Jensen's** and **AM-GM inequalities**. Let a,b,c be the side lengths of the triangle, and let s, R and S be the semiperimeter, circumradius and area respectively.*

The inequality can be written in the form

$$a(s-a) + b(s-b) + c(s-c) \geq 2\sqrt{3}S$$

*Applying **AM-GM inequality** and **Heron's formula**, we get*

$$a(s-a) + b(s-b) + c(s-c) \geq 3\sqrt[3]{abc(s-a)(s-b)(s-c)}$$

$$= 3S\sqrt[3]{\frac{4R}{s}}$$

*According to **Jensen's inequality**, we obtain*

$$\sin A + \sin B + \sin C \leq 3\sin\left(\frac{A+B+C}{3}\right)$$

$$= 3\sin\left(\frac{\pi}{3}\right)$$

$$= \frac{3\sqrt{3}}{2}$$

*By the **Sine law**, we derive*

$$a + b + c \leq 3\sqrt{3}R$$

which is equivalent to

$$s \leq \frac{3\sqrt{3}}{2}R$$

Therefore

$$a(s-a) + b(s-b) + c(s-c) \geq 3S\sqrt[3]{\frac{4R}{s}}$$

$$\geq 3S\sqrt[3]{\frac{8\sqrt{3}}{9}}$$

$$= 2\sqrt{3}S$$

as desired. Equality holds for $a = b = c$.

■

Solution 52.

*Using **Taylor's Theorem** with **Lagrange Remainder**, we get for all $x > 0$, there exists θ_x such that*

$$x^x = f(2) + \frac{f'(2)}{1!}(x-2) + \frac{f''(\theta_x)}{2!}(x-2)^2$$

On the other hand, we know that

$$\begin{cases} f'(x) = x^x(\ln x + 1) \\ f''(x) = x^x\left(\frac{1}{x} + (\ln x + 1)^2\right) > 0 \end{cases}$$

Thus

$$x^x \geq f(2) + \frac{f'(2)}{1!}(x-2)$$
$$= 4 + 4(1 + \ln 2)(x - 2)$$

Therefore

$$\sum_{cyc}(a+b)^{a+b} \geq \sum_{cyc}(4 + 4(1 + \ln 2)(a + b - 2))$$
$$= 4(a+b+c) + (4 + 8\ln 2)\sum_{cyc}(a-1)$$
$$= 4(a+b+c) + (4 + 8\ln 2)(a+b+c-3)$$
$$\geq 4(a+b+c) + (4 + 8\ln 2)(3\sqrt[3]{abc} - 3)$$
$$= 4(a+b+c)$$

The proof is complete. Equality holds for a = b = c = 1.

∎

Solution 53.

We will give three solutions to this problem.

• *First Solution:*

*According to **Theorem 42 (Belabess)**: For n = 3 and a, b, c, x, y, z ≥ 0, we have*
$$\sum_{cyc} x(a+b) \geq 2\sqrt{(x+y+z)(xab + ybc + zca)}$$

Therefore

$$\sum_{cyc}\frac{a^3 + b^3}{1 + ab} = \sum_{cyc} a^3 \left(\frac{1}{1+ab} + \frac{1}{1+ac}\right)$$
$$\geq 2\sqrt{(a^3 + b^3 + c^3)\left(\sum_{cyc}\frac{a^3}{(1+ab)(1+ac)}\right)}$$

It is sufficient to prove

$$\sum_{cyc}\frac{a^3}{(1+ab)(1+ac)} \geq \frac{3}{4}$$

*Applying **Cauchy-Schwarz inequality**, we find that*

$$\sum_{cyc}\frac{a^3}{(1+ab)(1+ac)} \geq \frac{\left(\sum_{cyc}a^2\right)^2}{\sum_{cyc}a(1+ab)(1+ac)}$$

Therefore, it suffices to prove that

$$\frac{\left(\sum_{cyc} a^2\right)^2}{\sum_{cyc} a(1+ab)(1+ac)} \geq \frac{3}{4}$$

Obviously $abc \leq 1$ *and according to* **Problem 1**, *we get*

$$\sum_{cyc} a^2 b \leq \sum_{cyc} a^2$$

Consequently

$$\sum_{cyc} a(1+ab)(1+ac) = \sum_{cyc}(a + a^2 b + a^2 c + a^3 bc)$$

$$\leq 3 + 3\left(\sum_{cyc} a^2\right)$$

Finally

$$4\left(\sum_{cyc} a^2\right)^2 - 3\sum_{cyc} a(1+ab)(1+ac) \geq 4\left(\sum_{cyc} a^2\right)^2 - 9\left(1+\sum_{cyc} a^2\right)$$

$$= \left(\sum_{cyc} a^2 - 3\right)\left(4\sum_{cyc} a^2 + 3\right)$$

$$\geq 0$$

The problem is completely solved. Equality holds for $a = b = c = 1$.

• **Second Solution:**

Applying **Cauchy Schwarz inequality**, *we deduce*

$$\sum_{cyc} \frac{a^3 + b^3}{1+ab} \geq \frac{\left(\sum_{cyc} \sqrt{a^3 + b^3}\right)^2}{3 + ab + bc + ca}$$

$$= \frac{2\left(\sum_{cyc} a^3\right) + 2\sum_{cyc}\sqrt{(a^3+b^3)(a^3+c^3)}}{3 + ab + bc + ca}$$

$$\geq \frac{4\left(\sum_{cyc} a^3\right) + 2\sum_{cyc}(ab)^{\frac{3}{2}}}{3 + ab + bc + ca}$$

$$= \frac{3\left(\sum_{cyc} a^3\right) + \left(\sum_{cyc} a^{\frac{3}{2}}\right)^2}{3 + ab + bc + ca}$$

$$\geq \frac{2\sqrt{3\left(\sum_{cyc} a^3\right)\left(\sum_{cyc} a^{\frac{3}{2}}\right)}}{3 + ab + bc + ca}$$

*Applying **Power-Mean inequality**, we get*

$$\sum_{cyc} a^{\frac{3}{2}} \geq 3$$

On the other hand, we know that

$$ab + bc + ca \leq \frac{(a+b+c)^2}{3} = 3$$

Therefore

$$\sum_{cyc} \frac{a^3 + b^3}{1 + ab} \geq \frac{2\sqrt{3 \left(\sum_{cyc} a^3\right) \left(\sum_{cyc} a^{\frac{3}{2}}\right)}}{3 + ab + bc + ca}$$

$$\geq \frac{2\sqrt{3 \left(\sum_{cyc} a^3\right) . 3}}{3 + 3}$$

$$= \sqrt{3(a^3 + b^3 + c^3)}$$

as required. Equality holds for $a = b = c = 1$.

- **Third Solution:**

WLOG, suppose that $a \geq b \geq c$. Thus, $c \leq 1$ and $abc \leq 1$.

Let

$$L = \frac{a^3 + b^3}{1 + ab} + \frac{b^3 + c^3}{1 + bc} + \frac{c^3 + a^3}{1 + ca} - \left(\frac{a^3 + b^3 + c^3}{1 + ab} + \frac{b^3}{1 + bc} + \frac{c^3}{1 + c^2} + \frac{a^3}{1 + ca}\right)$$

We have

$$L = \frac{c^3(a - c)(b - c)(1 - abc^2)}{(1 + ab)(1 + bc)(1 + ca)(c^2 + 1)} \geq 0$$

*Applying **AM-GM inequality**, we deduce*

$$\frac{a^3 + b^3 + c^3}{1 + ab} + \frac{b^3}{1 + bc} + \frac{c^3}{1 + c^2} + \frac{a^3}{1 + ca} \geq 2\sqrt{\left[\frac{a^3 + b^3 + c^3}{1 + ab}\right]\left[\frac{b^3}{1 + bc} + \frac{c^3}{1 + c^2} + \frac{a^3}{1 + ca}\right]}$$

Therefore, it suffices to show that

$$\frac{b^3}{1 + bc} + \frac{c^3}{1 + c^2} + \frac{a^3}{1 + ca} \geq \frac{3(1 + ab)}{4}$$

*According to **Hölder's inequality**, we get*

$$3\left[\frac{b^3}{1 + bc} + \frac{c^3}{1 + c^2} + \frac{a^3}{1 + ca}\right][1 + bc + 1 + c^2 + 1 + ac] \geq (a + b + c)^3$$

Consequently, we only need to prove

$$(1 + ab)(1 + c) \leq 4$$

*According to **AM-GM inequality**, it is enough to prove*

$$\left(1 + \frac{(a+b)^2}{4}\right)(1+c) \leq 4$$

which is equivalent to

$$\frac{(a+b)(1-c)^2}{4} \geq 0$$

The proof is complete.

∎

5.1.2 Non-symmetric inequalities

Solution 54.

First, notice that we can rewrite the inequality in the following form

$$\sum_{cyc} a(2a^2 + 3b^2) = \sum_{cyc} a(2a^2 + 3ac)$$

*Applying **Cauchy-Schwarz inequality**, we obtain*

$$\sum_{cyc} a(2a^2 + 3ac) \leq \sqrt{\sum_{cyc} a^2} \sqrt{\sum_{cyc} (2a^2 + 3ac)^2}$$

Furthermore

$$\sum_{cyc} (2a^2 + 3ac)^2 = \sum_{cyc} (4a^4 + 9a^2c^2 + 12a^3c)$$

$$= 4\sum_{cyc} (a^2 + 2a^2b^2) + \sum_{cyc} a^2b^2 + 12\left(\sum_{cyc} a^3b\right)$$

$$= 4(a^2 + b^2 + c^2)^2 + \sum_{cyc} a^2b^2 + 12\left(\sum_{cyc} a^3b\right)$$

On the other hand, we know that

$$\sum_{cyc} a^2b^2 \leq \frac{(a^2 + b^2 + c^2)^2}{3}$$

*Applying **Vasc's inequality**, we get*

$$\sum_{cyc} a^3b \leq \frac{(a^2 + b^2 + c^2)^2}{3}$$

We conclude that

$$\sum_{cyc} (2a^2 + 3ac)^2 = 4(a^2 + b^2 + c^2)^2 + \sum_{cyc} a^2b^2 + 12\sum_{cyc} a^3b$$

$$\leq 4(a^2 + b^2 + c^2)^2 + \frac{(a^2 + b^2 + c^2)^2}{3} + \frac{12(a^2 + b^2 + c^2)^2}{3}$$

$$= \frac{25(a^2 + b^2 + c^2)^2}{3}$$

Finally

$$\sum_{cyc} a(3a^2 + 2b^2) \leq \sqrt{\sum_{cyc} a^2} \sqrt{\sum_{cyc} (3a^2 + 2ac)^2}$$

$$\leq \sqrt{\frac{25(a^2 + b^2 + c^2)^3}{3}}$$

$$= 15$$

This ends the proof. The equality holds for $a = b = c = 1$.

Comment 18.

*We can prove **Vasc's inequality** in a variety of ways:*

○ **First proof:**

*We will be using the **Discriminant method**. WLOG, we can assume $a = \min\{a, b, c\}$ and let $x = b - a$ and $y = c - a$. The original inequality is then equivalent to*

$$f(a, x, y) = (x^2 + y^2 - xy)a^2 + (x^3 + y^3 + 4xy^2 - 5x^2y)a + x^4 + y^4 + 2x^2y^2 - 3x^3y$$

which is a quadratic polynomial on a. Therefore, it is enough to check that

$$\Delta \leq 0$$

On the other hand, we have

$$\Delta = (x^3 + y^3 + 4xy^2 - 5x^2y)^2 - 4(x^2 + y^2 - xy)(x^4 + y^4 + 2x^2y^2 - 3x^3y)$$
$$= -3(x^3 - x^2y - 2xy^2 + y^3)^2$$
$$\leq 0$$

Therefore

$$f(a, x, y) \geq 0$$

The proof is complete.

○ **Second proof:**

we are going to use the following well-known inequality

$$(x + y + z)^2 \geq 3(xy + yz + zx)$$

If we take $x = a^2 + bc - ab$, $y = b^2 + ac - bc$ and $z = c^2 + ab - ac$, we get

$$(a^2 + b^2 + c^2)^2 = \left(\sum_{cyc} (a^2 + bc - ab) \right)^2$$

$$= \left(\sum_{cyc} x \right)^2$$

$$\geq 3(xy + yz + zx)$$

$$= 3 \left(\sum_{cyc} (a^2 + bc - ab)(b^2 + ac - bc) \right)$$

$$= 3(a^3 b + b^3 c + c^3 a)$$

The proof is complete.

○ **Third proof:**

Vasc's inequality *can also be proved using the following identity*

$$(a^2 + b^2 + c^2)^2 - 3(a^3 b + b^3 c + c^3 a) = \frac{1}{6} \sum_{cyc} \left(a^2 - 3ab + b^2 - 2c^2 + 3ca \right)^2 \geq 0$$

∎

Solution 55.

*We will use the **Cauchy Reverse technique** to prove this inequality. In fact*

$$\frac{a}{b^2 + 1} = \frac{a(1 + b^2 - b^2)}{b^2 + 1}$$

$$= a - \frac{ab^2}{b^2 + 1}$$

$$\geq a - \frac{ab^2}{2b}$$

$$= a - \frac{ab}{2}$$

(Alternatively, notice that $\frac{a}{b^2+1} - \frac{2a-ab}{2} = \frac{ab(b-1)^2}{2(b^2+1)}$)

Therefore

$$\sum_{cyc} \frac{a}{b^2 + 1} \geq \sum_{cyc} \frac{2a - ab}{2}$$

$$= \frac{2(a + b + c) - ab - bc - ca}{2}$$

$$= \frac{12 - 2ab - 2bc - 2ca}{4}$$

$$= \frac{3 + (a + b + c)^2 - 2ab - 2bc - 2ca}{4}$$

$$= \frac{3 + a^2 + b^2 + c^2}{4}$$

$$\geq \frac{\sqrt{3(a^2 + b^2 + c^2)}}{2}$$

as required. Equality holds for $a = b = c = 1$.

Comment 19.

*There are times when proving inequalities with fractions in which applying the **AM-GM** inequality to each denominator results in the wrong sign for the resulting expression. We will show in the following examples how the **Cauchy Reverse technique** could be used to overcome this difficulty.*

○ ***First example:** [AoPS]*

Let x, y, z be positive real numbers. Prove that:

$$\frac{x^3}{x^2+y^2} + \frac{y^3}{y^2+z^2} + \frac{z^3}{z^2+x^2} \geq \frac{x+y+z}{2}$$

○ ***Proof:***

We know that $x^2 + y^2 \geq 2xy$, therefore

$$\sum_{cyc} \frac{x^3}{x^2+y^2} = \sum_{cyc} \frac{x(x^2+y^2-y^2)}{x^2+y^2}$$

$$= \sum_{cyc} \left(x - \frac{xy^2}{x^2+y^2} \right)$$

$$\geq \sum_{cyc} \left(x - \frac{xy^2}{2xy} \right)$$

$$= \sum_{cyc} \left(x - \frac{y}{2} \right)$$

$$= \frac{x+y+z}{2}$$

as required. Equality holds for $x = y = z$.

○ ***Second example:** [AoPS]*

Let a, b, c, d be positive real numbers such that $a^2 + b^2 + c^2 + d^2 = 4$. Prove that:

$$\frac{1}{a^3+2} + \frac{1}{b^3+2} + \frac{1}{c^3+2} + \frac{1}{d^3+2} \geq \frac{4}{3}$$

○ ***Proof:***

For any $x \geq 0$, we have $x^3 + 2 \geq 3x$. Therefore

$$\sum_{cyc} \frac{1}{a^3+2} = \frac{1}{2} \sum_{cyc} \left(1 - \frac{a^3}{a^3+2} \right)$$

$$\geq \frac{1}{2} \sum_{cyc} \left(1 - \frac{a^3}{3a} \right)$$

$$= \frac{1}{2} \sum_{cyc} \left(1 - \frac{a^2}{3} \right)$$

$$= \frac{4}{3}$$

as required. Equality at $a = b = c = d = 1$.

∎

Solution 56.

First of all, we will prove the following inequality

$$\frac{a}{b} + \frac{b}{c} + \frac{c}{a} \geq \frac{\sqrt{3(a^2 + b^2 + c^2)}}{\sqrt[3]{abc}}$$

For proving the lemma, we square the inequality and need to prove that

$$\frac{a^2}{b^2} + \frac{b^2}{c^2} + \frac{c^2}{a^2} + 2\left(\frac{a}{c} + \frac{b}{a} + \frac{c}{b}\right) \geq \frac{3(a^2 + b^2 + c^2)}{\sqrt[3]{(abc)^2}}$$

*Applying **AM-GM inequality**, we have*

$$\frac{a^2}{b^2} + \frac{2a}{c} \geq 3\sqrt[3]{\frac{a^4}{b^2 c^2}} = \frac{3a^2}{\sqrt[3]{(abc)^2}}$$

Making two similar inequalities and add all of them together to finish the proof.

Going back to the original problem, we get

$$\frac{a}{b} + \frac{b}{c} + \frac{c}{a} + abc \geq \frac{\sqrt{3(a^2 + b^2 + c^2)}}{\sqrt[3]{abc}} + abc$$

$$= \frac{3}{\sqrt[3]{abc}} + abc$$

$$\geq 4\sqrt[4]{\left(\frac{1}{\sqrt[3]{abc}}\right)^3 . abc}$$

$$= 4$$

as desired. Equality holds for $a = b = c = 1$.

∎

Solution 57.

We will give four solutions to this inequality.

• ***First Solution:***

WLOG, we can assume $a, b \geq c$. We have

$$a^3 + b^3 + c^3 - 3abc = (a + b + c)((a - b)^2 + (a - c)(a - c))$$

On the other hand

$$a + b + c \geq (a - c) + (b - c) \geq 2\sqrt{(a-b)(b-c)}$$

*Applying **AM-GM inequality**, we obtain*

$$(a - b)^2 + (a - c)(a - c) \geq 2|a - b|\sqrt{(a-b)(b-c)}$$

Finally

$$a^3 + b^3 + c^3 - 3abc \geq 4|(a-b)(b-c)(c-a)|$$

The proof is complete.

• *Second Solution:*

WLOG, we can assume $a \leq b, c$.

(i) **The first case: $b < c$.**

Therefore

$$(a - b)(b - c)(c - a) \geq 0$$

and the problem is obvious.

(ii) **The second case: $b \geq c$.**

*Using the **Buffalo Way technique**: let $x = b - c \geq 0$ and $y = c - a \geq 0$. Therefore, we can rewrite the inequality as follows*

$$3a(x^2 + xy + y^2) + x^3 - x^2 y - xy^2 + 2y^3$$

*Obviously, we have $x^2 + xy + y^2 \geq 0$ and using the **Rearrangement inequality**, we get*

$$x^3 - x^2 y - xy^2 + 2y^3 \geq x^3 - x^2 y - xy^2 + y^3$$
$$\geq 0$$

The proof is complete. Equality holds for $a = b = c$.

• *Third Solution:*

We will prove the stronger inequality

$$a^3 + b^3 + c^3 \geq \sqrt{9 + 6\sqrt{3}}(a-b)(b-c)(c-a) + 3abc$$

*Let $b = a + x, c = a + x + y$ and $t = \frac{x}{y}$ where $x, y > 0$ (For $x = 0$ or $y = 0$, the inequality is just **AM-GM inequality**).*

We get

$$(3a + 2x + y)(x^2 + xy + y^2) \geq \sqrt{9 + 6\sqrt{3}}xy(x+y)$$

Thus, it is sufficient to prove

$$(2x + y)(x^2 + xy + y^2) \geq \sqrt{9 + 6\sqrt{3}} xy(x + y)$$

which equivalent to

$$(2t + 1)(t^2 + t + 1) \geq \sqrt{9 + 6\sqrt{3}} t(t + 1)$$

This is a polynomial P of degree 3 that can be factorised as follows (at the minimum, we should have $P(t) = 0$ and $P'(t) = 0$)

$$\left(2t - \sqrt{3 + 2\sqrt{3}} + 1\right)^2 \left(2t + 1 + \sqrt{2\sqrt{3} - 3}\right) \geq 0$$

which is true with equality at

$$t = \frac{1}{2}\left(\sqrt{3 + 2\sqrt{3}} - 1\right)$$

The problem is completely solved.

• *Fourth Solution:*

*We will use the **Entirely Mixing Variables method**. The above inequality can be rewritten as follows*

$$(a + b + c)((a - b)^2 + (b - c)^2 + (c - a)^2) \geq 8(a - b)(b - c)(c - a)$$

WLOG, we may assume that $c = \min(a, b, c)$. Fix the differences $a - b$, $b - c$ and $c - a$ and decrease simultaneously from a, b, c one value c (which means that we supersede a, b, c by $a - c, b - c, 0$). Therefore, $a - b$, $a - c$, $b - c$ don't change but $a + b + c$ is decreased.

The left hand side of the inequality is decreased but the right hand side is invariable. Consequently, we only need to prove the problem in case $a, b \geq c = 0$, the problem becomes

$$a^3 + b^3 \geq 4ab(b - a)$$

This inequality is obviously true because

$$a^3 + b^3 - 4ab(b - a) = a^3 + b(b - 2a)^2 \geq 0$$

This ends the proof. The equality occurs if and only if $a = b = c$.

■

Solution 58.

We will give two solutions to this problem.

• *First Solution:*

The given constraint tells that it suffices to prove

$$\frac{(a+b)(b+c)(c+a)}{abc} \geq \frac{24(a^2+b^2+c^2)}{(a+b+c)^2}$$

Since the LHS is a decreasing function of $abc = w^3$, by uvw method it is enough to check the case $b = c = 1$, and we get

$$\frac{(a+1)^2}{a} \geq \frac{12(a^2+2)}{(a+2)^2}$$

which is equivalent to

$$(a-1)^2(a-2)^2 \geq 0$$

We are done. Equality holds for (a, a, a), $(2a, a, a)$ and permutations.

• *Second Solution:*

It is enough to prove

$$\frac{(a+b)(b+c)(c+a)}{abc} \geq \frac{24(a^2+b^2+c^2)}{(a+b+c)^2}$$

We have

$$\frac{(a+b)(b+c)(c+a)}{abc} - \frac{24(a^2+b^2+c^2)}{(a+b+c)^2} = \sum_{cyc} \frac{(a+b-3c)^2(a-b)^2}{ab(a+b+c)^2} \geq 0$$

as required. Equality holds for (a, a, a), $(2a, a, a)$ and permutations.

∎

Solution 59.

The inequality is equivalent to prove

$$\sum_{cyc}\left(\frac{a}{b} - \frac{a}{b+c}\right) \geq \frac{3}{2} \iff \sum_{cyc} \frac{ac}{b(b+c)} \geq \frac{3}{2}$$

According to Cauchy-Schwarz inequality, we find that

$$\sum_{cyc} \frac{ac}{b(b+c)} \geq \frac{\left(\sqrt{\frac{ac}{b}} + \sqrt{\frac{ab}{c}} + \sqrt{\frac{bc}{a}}\right)^2}{2(a+b+c)}$$

Using the well-known inequality $(x+y+z)^2 \geq 3(xy+yz+zx)$, we get

$$\left(\sqrt{\frac{ac}{b}} + \sqrt{\frac{ab}{c}} + \sqrt{\frac{bc}{a}}\right)^2 \geq 3(a+b+c)$$

Finally

$$\sum_{cyc} \frac{ac}{b(b+c)} \geq \frac{3(a+b+c)}{2(a+b+c)} = \frac{3}{2}$$

The proof is complete. Equality holds for $a = b = c$.

∎

Solution 60.

We have $abc = 1$, so there exist $x, y, z > 0$ such that $a = \frac{y}{z}$, $b = \frac{z}{x}$ and $c = \frac{x}{y}$. Therefore, the inequality is equivalent to

$$5(x^3y^3 + y^3z^3 + z^3x^3) + 12x^2y^2z^2 \geq 9xyz(xy^2 + yz^2 + zx^2)$$

Let $a_1 = yz$, $b_1 = zx$ and $c_1 = xy$. We get

$$5(a_1^3 + b_1^3 + c_1^3) + 12a_1b_1c_1 \geq 9(a_1^2b_1 + b_1^2c_1 + c_1^2a_1)$$

We will prove this last inequality using the **Buffalo Way technique**. WLOG, we can assume $a_1 \leq b_1, c_1$ and let $u = b_1 - a_1$, $v = c_1 - a_1$. The inequality is equivalent to

$$6(u^2 - uv + v^2)a_1 + 5u^3 - 9u^2v + 5v^3 \geq 0$$

Obviously
$$u^2 - uv + v^2 \geq 0$$

Therefore, we only need to prove

$$5u^3 - 9u^2v + 5v^3 \geq 0$$

Applying **AM-GM inequality**

$$\frac{5u^3}{2} + \frac{5u^3}{2} + 5v^3 \geq 3\sqrt[3]{\frac{125}{4}u^2v} \geq 9u^2v$$

The proof is complete. Equality holds for $a = b = c = 1$.

∎

Solution 61.

Applying **Titu's lemma**, we obtain

$$\sum_{cyc} \frac{a}{b+1} \geq \frac{(a+b+c)^2}{a+b+c+ab+bc+ca}$$

Thus, it is enough to prove

$$2\left(\sum_{cyc} a^2\right) + \sum_{cyc} ab \geq 3\left(\sum_{cyc} a\right)$$

which is true because

$$2\left(\sum_{cyc} a^2\right) + \sum_{cyc} ab = \frac{(a+b+c)^2}{2} + \frac{9}{2} \geq 3(a+b+c)$$

as desired. Equality holds for $a = b = c = 1$.

■

Solution 62.

*We will use **Cauchy Reverse technique** to prove this inequality.*

Let $a^2 + b^2 + c^2 = k(ab + bc + ca)$. *Therefore* $k \geq 1$ *and we have*

$$\sum_{cyc} \frac{a}{b^2+1} - \frac{3}{2} = \sum_{cyc}\left(\frac{a}{b^2+1} - a\right) + a + b + c - \frac{3}{2}$$

$$= \sum_{cyc} \frac{-ab^2}{b^2+1} + a + b + c - \frac{3}{2}$$

Using $x^2 + 1 \geq 2x$, *we get*

$$\sum_{cyc} \frac{a}{b^2+1} - \frac{3}{2} \geq \sum_{cyc} \frac{-ab^2}{2b} + a + b + c - \frac{3}{2}$$

$$= \sum_{cyc} \frac{-ab}{2} + a + b + c - \frac{3}{2}$$

$$= \sqrt{\frac{(a+b+c)^2(a^2+b^2+c^2)}{3}} - \frac{1}{2}\sum_{cyc}(a^2+ab)$$

$$= \frac{ab+bc+ca}{2\sqrt{3}}\left(2\sqrt{(k+2)k} - \sqrt{3}(k+1)\right)$$

$$= \frac{(ab+bc+ca)(k-1)(k+3)}{2\sqrt{3}\left(2\sqrt{(k+2)k} + \sqrt{3}(k+1)\right)}$$

$$\geq 0$$

Equality at $k = 1$ *meaning that* $a = b = c = 1$.

■

Solution 63.

Let $\frac{a}{b} = x$, $\frac{b}{c} = y$, $\frac{c}{a} = z$. *The inequality becomes*

$$2 + \sum_{sym} \frac{a^2}{b^2} \geq 4\left(\sum_{cyc} \frac{a}{b} - 1\right)$$

which is equivalent to

$$6 + \sum_{cyc} x^2 + \sum_{cyc} \frac{1}{x^2} \geq 4(x + y + z)$$

Let $p = x + y + z$, $q = xy + yz + zx$ *and* $r = xyz = 1$. *It is enough to prove*

$$p^2 - 6p + q^2 - 2q + 6 \geq 0$$

Applying **AM-GM inequality**, *we deduce*

$$\begin{aligned} p^2 - 6p + q^2 - 2q + 6 &\geq 2pq - 6p - 2q + 6 \\ &= 2(p-1)(q-3) \\ &\geq 0 \end{aligned}$$

where the last inequality is true as $p \geq 3$ *and* $q \geq 3$. *Equality holds for* $a = b = c = 1$.

∎

Solution 64.

It is enough to prove the two following results

$$\sum_{cyc} \frac{a}{b^2 + 1} \geq \frac{3\sqrt{3}}{4}$$

$$\sum_{cyc} \frac{1}{b^2 + 1} \geq \frac{3}{4}$$

○ *First inequality*

Equivalently, we show that if $ab + bc + ca = 1$, *then*

$$\sum_{cyc} \frac{b^2}{a(b^2 + 1)} = \sum_{cyc} \frac{b^2}{a(b + a)(b + c)} \geq \frac{3\sqrt{3}}{4}$$

Let $x = a + b + c$. *Applying* **Hölder's inequality**, *we get*

$$\begin{aligned} \sum_{cyc} \frac{b^2}{a(b + a)(b + c)} &\geq \frac{\left(\sum_{cyc} a\right)^3}{\left(\sum_{cyc}(b^2 + ab)\right)\left(\sum_{cyc}(ab + ac)\right)} \\ &= \frac{x^3}{2(x^2 - 1)} \\ &= \frac{3\sqrt{3}}{4} + \frac{(x - \sqrt{3})^2(2x + \sqrt{3})}{4(x^2 - 1)} \\ &\geq \frac{3\sqrt{3}}{4} \end{aligned}$$

as required.

○ **Second inequality**

Equivalently, we show that if $ab + bc + ca = 1$, then

$$\sum_{cyc} \frac{b^2}{b^2 + 1} \geq \frac{3}{4}$$

that we can easily prove using **Titu's lemma** as follows

$$\sum_{cyc} \frac{b^2}{b^2 + 1} \geq \frac{(a + b + c)^2}{a^2 + b^2 + c^2 + 3} = \frac{(a + b + c)^2}{(a + b + c)^2 + 1} \geq \frac{3}{4}$$

where the last step is true because $a + b + c \geq \sqrt{3(ab + bc + ca)} = \sqrt{3}$.

■

Solution 65.

We will present two solutions to this problem.

• First Solution

According to **Theorem 42 (Belabess)**, we get

$$\frac{a}{b} + \frac{b}{c} + \frac{c}{a} + 3 = \frac{a + b}{b} + \frac{b + c}{c} + \frac{c + a}{a}$$

$$\geq 2\sqrt{(a + b + c)\left(\frac{1}{a} + \frac{1}{b} + \frac{1}{c}\right)}$$

Let $t = \sqrt{(a + b + c)\left(\frac{1}{a} + \frac{1}{b} + \frac{1}{c}\right)}$. Therefore

$$\frac{a}{b} + \frac{b}{c} + \frac{c}{a} \geq 2t - 3$$

According to **AM-GM inequality**, we get

$$t \geq \sqrt{3\sqrt[3]{abc}.3\sqrt[3]{\frac{1}{abc}}} = 3$$

Consequently, it is enough to prove

$$5(2t - 3)^2 \geq 2t^2 + 27$$

that we can rewrite as follows

$$6(t - 3)(3t - 1) \geq 0$$

which is obviously true for $t \geq 3$. Equality holds for $a = b = c$.

● *Second Solution*

The original inequality is equivalent to

$$5 \left(\sum_{cyc} \frac{a^2}{b^2} \right) + 8 \left(\sum_{cyc} \frac{b}{a} \right) \geq 33 + 2 \left(\sum_{cyc} \frac{a}{b} \right)$$

*Applying **AM-GM inequality**, we obtain*

$$4 \left(\sum_{cyc} \frac{a^2}{b^2} \right) + 7 \left(\sum_{cyc} \frac{b}{a} \right) \geq 4 \cdot 3 + 7 \cdot 3 = 33$$

Consequently, it is enough to prove

$$\sum_{cyc} \frac{a^2}{b^2} + \sum_{cyc} \frac{b}{a} \geq 2 \left(\sum_{cyc} \frac{a}{b} \right)$$

which can be done as follows

$$\sum_{cyc} \frac{a^2}{b^2} + \sum_{cyc} \frac{b}{a} = \frac{1}{2} \left(\sum_{cyc} \frac{a^2}{b^2} \right) + \frac{1}{2} \left(\sum_{cyc} \frac{a}{b} \right)^2$$

$$\geq \frac{2}{3} \left(\sum_{cyc} \frac{a}{b} \right)^2$$

$$\geq 2 \left(\sum_{cyc} \frac{a}{b} \right)$$

This ends the proof. Equality holds for $a = b = c$.

∎

Solution 66.

Notice that

$$\frac{a+b}{b+c} + \frac{b+c}{c+a} + \frac{c+a}{a+b} - 3 = \frac{(a-b)^2}{(a+c)(b+c)} + \frac{(a-c)(b-c)}{(a+b)(b+c)}$$

$$\frac{a^2+b^2+c^2+9}{4} - 3 = \frac{3\left(a^2+b^2+c^2\right) - (a+b+c)^2}{12} = \frac{(a-b)^2 + (a-c)(b-c)}{6}$$

The original inequality is equivalent to prove

$$L(a,b,c) = Q.(a-b)^2 + T.(a-c)(b-c) \geq 0$$

with

$$\begin{cases} Q = \frac{2(a+b+c)^2}{3(a+c)(b+c)} - 1 \\ T = \frac{2(a+b+c)^2}{3(a+b)(b+c)} - 1 \end{cases}$$

We have

$$L(a,b,c) = \left(Q - \frac{2(a-c)(b-c)}{3(a+c)(b+c)}\right)(a-b)^2 + \left(T + \frac{2(a-b)^2}{3(a+c)(b+c)}\right)(a-c)(b-c)$$

Assume that $c = \min(a,b,c)$. We have

$$Q - \frac{2(a-c)(b-c)}{3(a+c)(b+c)} = \frac{\left[\left(\sum_{cyc} a\right)^2 - 3\left(\sum_{cyc} ab\right)\right] + (a+b+c)^2 + 2c(a+b) - 5c^2 - 2ab}{3(a+c)(b+c)}$$

$$\geq \frac{(a+b+c)^2 - 5c^2 + 4c^2 - 2ab}{3(a+c)(b+c)}$$

$$> 0$$

Moreover

$$T + \frac{2(a-b)^2}{3(a+c)(b+c)} = \frac{2(a+b+c)^2}{3(a+b)(b+c)} + \frac{2(a-b)^2}{3(a+c)(b+c)} - 1$$

$$\geq \frac{2(a+b+c)^2 + 2(a-b)^2}{3(a+b)(b+c)} - 1$$

$$= \frac{(2a-b)^2 + 2c^2 + ab + bc + ca}{3(a+b)(b+c)}$$

$$\geq 0$$

as desired. Equality holds for $a = b = c = 1$.

■

Solution 67.

We will present two solutions to this problem.

● *First Solution*

*According to **Theorem 42 (Belabess)**, we get*

$$\frac{a}{b} + \frac{b}{c} + \frac{c}{a} + 3 = \frac{a+b}{b} + \frac{b+c}{c} + \frac{c+a}{a}$$

$$\geq 2\sqrt{\left(\frac{1}{a} + \frac{1}{b} + \frac{1}{c}\right)(a+b+c)}$$

$$= 2(a+b+c)$$

This ends the proof. Equality holds for $a = b = c = 1$.

● *Second Solution*

*We will apply **Aczel's inequality**: for $A \geq B \geq 0$ and $X \geq Y \geq 0$*

$$AX - BY \geq \sqrt{A^2 - B^2}\sqrt{X^2 - Y^2}$$

that we can easily prove

$$(AX - BY)^2 - (A^2 - B^2)(X^2 - Y^2) = (AY - BX)^2$$

Let

$$L = \frac{a}{b} + \frac{b}{c} + \frac{c}{a} + 3$$

Using this above-mentioned lemma

$$
\begin{aligned}
L &= \frac{a}{b} + \frac{b}{c} + \frac{c}{a} + 3 \\
&= \frac{ab^2 + bc^2 + ca^2 + 3abc}{abc} \\
&= \frac{(a+b+c)(ab+bc+ca) - (a^2b + b^2c + c^2a)}{abc} \\
&\geq \frac{(a+b+c)(ab+bc+ca) - \sqrt{a^2+b^2+c^2}.\sqrt{a^2b^2+b^2c^2+c^2a^2}}{abc} \\
&\geq \frac{\sqrt{(a+b+c)^2 - (\sqrt{a^2+b^2+c^2})^2}.\sqrt{(ab+bc+ca)^2 - (\sqrt{a^2b^2+b^2c^2+c^2a^2})^2}}{abc} \\
&\geq \frac{\sqrt{4abc(a+b+c)(ab+bc+ca)}}{abc} \\
&= \frac{2(ab+bc+ca)}{abc} \\
&= 2(a+b+c)
\end{aligned}
$$

The proof is complete.

■

Solution 68.

There are two cases to consider:

(i) **The first case:** $a \geq b \geq c$.

> *We have*
> $$LHS = \sum_{cyc} \frac{a}{b} - 3 \geq 0 \geq 2(a-b)(b-c)(c-a)$$

(ii) **The second case:** $a \leq b \leq c$.

> *Applying **AM-GM inequality**, we get*
> $$2(b-a)ac \leq \left(\frac{b-a+2a+c}{3}\right)^3 = 1$$

> *Moreover*
> $$LHS = \sum_{cyc} \frac{a}{b} - 3$$

$$= \frac{(a-b)^2}{ab} + \frac{(c-a)(c-b)}{ac}$$
$$\geq \frac{(c-a)(c-b)}{ac}$$
$$\geq 2(b-a)(c-a)(c-b)$$
$$= 2(a-b)(b-c)(c-a)$$

as desired.

∎

Solution 69.

○ *Left inequality:*

*According to **Cauchy-Schwarz inequality**, we deduce*

$$\sqrt{a}.b + \sqrt{b}.c + \sqrt{c}.a \leq \sqrt{(a+b+c)(ab+bc+ca)}$$
$$\leq \sqrt{\frac{(a+b+c)^3}{3}}$$
$$= \frac{1}{\sqrt{3}}$$

In a similar way, we get

$$2\sqrt{a}.c + 2\sqrt{b}.a + 2\sqrt{c}.b \leq 2\sqrt{(a+b+c)(ab+bc+ca)}$$
$$\leq 2\sqrt{\frac{(a+b+c)^3}{3}}$$
$$= \frac{2}{\sqrt{3}}$$

*According to **Power Mean inequality** (also by **Jensen's inequality**), we obtain*

$$\sqrt{a} + \sqrt{b} + \sqrt{c} \leq 3\sqrt{\frac{a+b+c}{3}} = \sqrt{3}$$

as desired. Equality holds for $\left(\frac{1}{3}, \frac{1}{3}, \frac{1}{3}\right)$.

○ *Right inequality:*

$$\sum_{cyc} \sqrt{a(b+2c+1)} \geq \sqrt{a} + \sqrt{b} + \sqrt{c}$$

$$= \sqrt{a+b+c+2\sqrt{ab}+\sqrt{bc}+\sqrt{ca}}$$
$$\geq \sqrt{a+b+c}$$
$$= 1$$

as desired. Equality holds for $(1,0,0)$ and permutations.

∎

Solution 70.

○ *Maximum:*

*By **AM-GM inequality** (also by **Rearrangement inequality**)*

$$\sum_{cyc} \left(x^3 + x^3 + y^3\right) \geq \sum_{cyc} 3x^2 y$$

Therefore

$$x^3 + y^3 + z^3 \geq x^2 y + y^2 z + z^2 x$$

Replace $a = x^2, b = y^2, c = z^2$, *then*

$$a\sqrt{a} + b\sqrt{b} + c\sqrt{c} \geq a\sqrt{b} + b\sqrt{c} + c\sqrt{a}$$

Or

$$(\sqrt{a} + \sqrt{b} + \sqrt{c})(a + b + c) \geq \sqrt{a}(b + 2c) + \sqrt{b}(c + 2a) + \sqrt{c}(a + 2b)$$

which implies that

$$0 \geq \sqrt{a}(b + 2c - 1) + \sqrt{b}(c + 2a - 1) + \sqrt{c}(a + 2b - 1)$$

hence the maximum is 0.

○ *Minimum:*

With $x, y, z \geq 0$, *we have*

$$(x^2 + y^2 + z^2)^3 \geq (x^3 + y^3 + z^3)^2$$

which is true since it is equivalent to

$$2\left(\sum_{cyc} x^2 y^2 (x^2 + y^2)\right) + \sum_{cyc} x^2 y^2 (x - y)^2 + 6x^2 y^2 z^2 \geq 0$$

Replace $a = x^2, b = y^2, c = z^2$ *to get*

$$1 = (a + b + c)^{\frac{3}{2}} \geq a\sqrt{a} + b\sqrt{b} + c\sqrt{c}$$

Thus

$$a\sqrt{b} + b\sqrt{c} + c\sqrt{a} + 1 - a\sqrt{a} - b\sqrt{b} - c\sqrt{c} \geq 0$$

which implies

$$\sqrt{a}(c - a) + \sqrt{b}(a - b) + \sqrt{c}(b - c) \geq -1$$

Or

$$\sqrt{a}(b + 2c - 1) + \sqrt{b}(c + 2a - 1) + \sqrt{c}(a + 2b - 1) \geq -1$$

Therefore the minimum is -1, *attained when* $a = b = 0, c = 1$ *and permutations.*

∎

Solution 71.

First of all, we will prove that

$$\frac{a}{a+b} + \frac{b}{b+c} + \frac{c}{c+a} \geq \frac{a+b+c}{a+b+c-\sqrt[3]{abc}}$$

For that, we will consider two cases:

(i) **The first case:** $ab + bc + ca \geq (a+b+c)\sqrt[3]{abc}$.

$$\frac{a}{a+b} + \frac{b}{b+c} + \frac{c}{c+a} = \sum_{cyc} \frac{a^2}{a^2 + ab}$$

$$\geq \frac{(a+b+c)^2}{\sum_{cyc}(a^2+ab)}$$

$$= \frac{(a+b+c)^2}{(a+b+c)^2 - ab - bc - ca}$$

$$\geq \frac{(a+b+c)^2}{(a+b+c)^2 - (a+b+c)\sqrt[3]{abc}}$$

$$= \frac{a+b+c}{a+b+c-\sqrt[3]{abc}}$$

(ii) **The second case:** $ab + bc + ca \leq (a+b+c)\sqrt[3]{abc}$.

$$\frac{a}{a+b} + \frac{b}{b+c} + \frac{c}{c+a} = \sum_{cyc} \frac{a^2c^2}{a^2c^2 + c^2ab}$$

$$\geq \frac{(ab+bc+ca)^2}{\sum_{cyc}(a^2c^2 + c^2ab)}$$

$$= \frac{(ab+bc+ca)^2}{(ab+bc+ca)^2 - abc(a+b+c)}$$

$$= \frac{(ab+bc+ca)^2}{(ab+bc+ca)^2 - \sqrt[3]{abc}\left(\sqrt[3]{abc}\right)^2(a+b+c)}$$

$$\geq \frac{(ab+bc+ca)^2}{(ab+bc+ca)^2 - \sqrt[3]{abc}\left(\frac{ab+bc+ca}{a+b+c}\right)^2(a+b+c)}$$

$$= \frac{a+b+c}{a+b+c-\sqrt[3]{abc}}$$

Let $t = \frac{a+b+c}{\sqrt[3]{abc}} \geq 3$. *Going back to the original inequality*

$$\frac{a}{a+b} + \frac{b}{b+c} + \frac{c}{c+a} + \frac{a+b+c}{4\sqrt[3]{abc}} - \frac{9}{4} \geq \frac{a+b+c}{a+b+c-\sqrt[3]{abc}} + \frac{a+b+c}{4\sqrt[3]{abc}} - \frac{9}{4}$$

$$= \frac{t}{t-1} + \frac{t}{4} - \frac{9}{4}$$

$$= \frac{(t-3)^2}{4(t-1)}$$

$$\geq 0$$

The proof is complete.

∎

Solution 72.

*Applying **Vasc's inequality**, we obtain*

$$(a^3 + b^3 + c^3)^2 \geq 3 \left(a^{\frac{9}{2}} b^{\frac{3}{2}} + b^{\frac{9}{2}} c^{\frac{3}{2}} + c^{\frac{9}{2}} a^{\frac{3}{2}} \right)$$

Therefore

$$(a^3 + b^3 + c^3)^5 \geq 9 \left(a^{\frac{9}{2}} b^{\frac{3}{2}} + b^{\frac{9}{2}} c^{\frac{3}{2}} + c^{\frac{9}{2}} a^{\frac{3}{2}} \right)^2 (a^3 + b^3 + c^3)$$

$$\geq 9 \left(a^4 b + b^4 c + c^4 a \right)$$

*where the last inequality is true by **Hölder's inequality**.*

∎

Solution 73.

Notice that

$$\sum_{cyc} \frac{a^2}{a + b} = \sum_{cyc} \frac{b^2}{a + b}$$

Therefore, we can rewrite the original inequality as

$$\sum_{cyc} \frac{a^2 + b^2}{a + b} + 2 \left(\sum_{cyc} \frac{1}{a + b} \right) \geq 2\sqrt{3(a + b + c)}$$

Or

$$\sum_{cyc} \left(a + b - \frac{2ab}{a + b} \right) + 2 \left(\sum_{cyc} \frac{1}{a + b} \right) \geq 2\sqrt{3(a + b + c)}$$

Or

$$a + b + c + \frac{1 - ab}{a + b} + \frac{1 - bc}{b + c} + \frac{1 - ca}{c + a} \geq \sqrt{3(a + b + c)}$$

Let $p = a + b + c$, $q = ab + bc + ca$ and $r = abc = 1$. We get

$$\sum_{cyc} \frac{1 - ab}{a + b} = \frac{p^2 - q^2 + q - p}{pq - 1}$$

Consequently, we need to prove

$$p + \frac{p^2 - q^2 + q - p}{pq - 1} \geq \sqrt{3p}$$

Let

$$f(q) = p + \frac{p^2 - q^2 + q - p}{pq - 1} - \sqrt{3p}$$

We will prove that f is a decreasing function on $[3, +\infty[$. For $q_2 \geq q_1 \geq 3$, we have

$$f(q_2) - f(q_1) = \frac{(q_1 - q_2)(p^3(p-1) + (p-1)q_1q_2 + (q_1-1)(q_2-1))}{(pq_2 - 1)(pq_1 - 1)} \leq 0$$

which means that f is decreasing for $q \geq 3$.

On the other hand, we know that

$$a^2 + b^2 + c^2 + 2abc + 1 \geq 2(ab + bc + ca)$$

Therefore

$$q \leq \frac{p^2 + 3}{4}$$

It is sufficient to prove

$$f\left(\frac{p^2 + 3}{4}\right) \geq 0$$

Let $x = \sqrt{\frac{p}{3}}$. We have $x \geq 1$ and

$$f\left(\frac{p^2 + 3}{4}\right) = \frac{3(x-1)(x(27x^4 - 9x^3 - 12x + 17) + 1)}{4(9x^4 + 3x^2 + 4)} \geq 0$$

which is obviously true.

■

Solution 74.

We need to prove

$$2 + \sum_{sym} \frac{a^2}{b^2} \geq 4\left(\sum_{cyc} \frac{a}{b} - 1\right)$$

Let $\frac{a}{b} = x$, $\frac{b}{c} = y$ and $\frac{c}{a} = z$. The inequality becomes

$$6 + \sum_{cyc} x^2 + \sum_{cyc} \frac{1}{x^2} \geq 4(x + y + z)$$

Denote $p = x + y + z$, $q = xy + yz + zx$ and $r = xyz = 1$. We need to prove

$$p^2 - 6p + q^2 - 2q + 6 \geq 0$$

Since $q \geq 3$, we have

$$p^2 - 6p + q^2 - 2q + 6 \geq p^2 - 6p + q(q - 2) + 6$$
$$\geq p^2 - 6p + 9$$
$$= (p - 3)^2$$
$$\geq 0$$

The proof is complete. Equality holds for $a = b = c = 1$.

Solution 75.

*Applying **AM-GM inequality**, we have for $x \geq 0$*

$$x^2 + x + 1 \geq 3x$$

It is sufficient to prove

$$\frac{a}{b} + \frac{b}{c} + \frac{c}{a} \geq \sqrt{3(a^2 + b^2 + c^2)}$$

For proving the lemma, we square the inequality and need to prove that

$$\frac{a^2}{b^2} + \frac{b^2}{c^2} + \frac{c^2}{a^2} + 2\left(\frac{a}{c} + \frac{b}{a} + \frac{c}{b}\right) \geq 3(a^2 + b^2 + c^2)$$

*By **AM-GM inequality**, we get*

$$\frac{a^2}{b^2} + 2\frac{a}{c} \geq 3\sqrt[3]{\frac{a^4}{b^2c^2}} = 3a^2$$

In a similar way, we obtain

$$\frac{b^2}{c^2} + 2\frac{b}{a} \geq 3\sqrt[3]{\frac{b^4}{c^2a^2}} = 3b^2$$

$$\frac{c^2}{a^2} + 2\frac{c}{b} \geq 3\sqrt[3]{\frac{c^4}{a^2b^2}} = 3c^2$$

Add all the three together to get the desired result. Equality holds for $a = b = c = 1$.

■

Solution 76.

We will use the following lemma to prove the inequality:

○ *Lemma:*

Given x, y, z are non-negative real numbers, then:

$$(x + y + z)^3 \geq \frac{27}{4}\left(x^2y + y^2z + z^2x + xyz\right)$$

○ *Proof:*

WLOG, we can assume that $x \geq y \geq z$ or $z \geq y \geq x$. Therefore

$$z(x - y)(y - z) \geq 0$$

Or

$$x^2y + yz^2 + 2xyz \geq x^2y + y^2z + z^2x + xyz$$

*Applying **AM-GM inequality**, we obtain*

$$x^2y + y^2z + z^2x + xyz \leq x^2y + yz^2 + 2xyz$$

$$= y(z+x)^2$$

$$\leq \frac{4}{27}\left(y + \frac{x+z}{2} + \frac{x+z}{2}\right)^3$$

$$= \frac{4}{27}(x+y+z)^3$$

The proof is complete.

○ **Application:**

According to the previous lemma, we deduce

$$\left(\frac{a}{b} + \frac{b}{c} + \frac{c}{a}\right)^3 \geq \frac{27}{4}\left(\frac{a^2}{bc} + \frac{b^2}{ca} + \frac{c^2}{ab} + 1\right)$$

$$= \frac{27}{4}\left(\frac{a^3 + b^3 + c^3}{abc} + 1\right)$$

It suffices to prove

$$\frac{a^3 + b^3 + c^3}{abc} + 1 \geq \frac{108(a^2 + b^2 + c^2)^3}{(a+b+c)^6}$$

Notice that

$$\frac{a^3 + b^3 + c^3}{abc} + 1 = \frac{(a+b+c)(a^2+b^2+c^2 - ab - bc - ca)}{abc} + 4$$

Since $(ab+bc+ca)^2 \geq 3abc(a+b+c)$, *we get*

$$\frac{a+b+c}{abc} \geq \frac{3(a+b+c)^2}{(ab+bc+ca)^2}$$

Therefore, we only need to prove

$$\frac{3(a+b+c)^2(a^2+b^2+c^2 - ab - bc - ca)}{(ab+bc+ca)^2} + 4 \geq \frac{108(a^2+b^2+c^2)^3}{(a+b+c)^6}$$

Let $t = \frac{3(a^2+b^2+c^2)}{(a+b+c)^2}$. *We have* $1 \leq t \leq 3$, *and the desired inequality now becomes*

$$\frac{54(t-1)}{3-t^2} + 4 \geq 4t^3$$

which is equivalent to

$$(t-1)\left[2(3+3t-t^2)(t-1)^2 + 3\right] \geq 0$$

This last inequality is true since $1 \leq t \leq 3$. *Equality holds for* $a = b = c = 1$.

■

Solution 77.

We will give two solutions to this problem.

- *First Solution:*

Let

$$L = \sqrt{7a^2 + b + \frac{1}{c}} + \sqrt{7b^2 + c + \frac{1}{a}} + \sqrt{7c^2 + a + \frac{1}{b}}$$

Applying **Generalised Minkowski's inequality,** *we obtain*

$$L \geq \sqrt{7(a+b+c)^2 + (\sqrt{a} + \sqrt{b} + \sqrt{c})^2 + \left(\frac{1}{\sqrt{a}} + \frac{1}{\sqrt{b}} + \frac{1}{\sqrt{c}}\right)^2}$$

$$\geq \sqrt{7(a+b+c)^2 + 2\left(\sqrt{a} + \sqrt{b} + \sqrt{c}\right)\left(\frac{1}{\sqrt{a}} + \frac{1}{\sqrt{b}} + \frac{1}{\sqrt{c}}\right)}$$

$$\geq \sqrt{21(ab + bc + ca) + 18}$$

$$= 9$$

as required. Equality holds for $a = b = c = 1$.

- *Second Solution:*

Let

$$L = \sqrt{7a^2 + b + \frac{1}{c}} + \sqrt{7b^2 + c + \frac{1}{a}} + \sqrt{7c^2 + a + \frac{1}{b}}$$

We have

$$L^2 = \left(\sum_{cyc} \sqrt{7a^2 + b + \frac{1}{c}}\right)^2$$

$$= \sum_{cyc}\left(7a^2 + a + \frac{1}{a} + 2\sqrt{\left(7a^2 + b + \frac{1}{c}\right)\left(7b^2 + c + \frac{1}{a}\right)}\right)$$

$$\geq \sum_{cyc}\left(7a^2 + a + \frac{1}{a} + 2\left(7ab + \sqrt{ab} + \frac{1}{\sqrt{ab}}\right)\right)$$

$$\geq \sum_{cyc}\left(7a^2 + 2\sqrt{a.\frac{1}{a}} + 2\left(7ab + 2\sqrt{\sqrt{ab}.\frac{1}{\sqrt{ab}}}\right)\right)$$

$$\geq \sum_{cyc}\left(7 + 2 + 2\left(7 + 2\right)\right)$$

$$= 81$$

Therefore

$$L = \sqrt{7a^2 + b + \frac{1}{c}} + \sqrt{7b^2 + c + \frac{1}{a}} + \sqrt{7c^2 + a + \frac{1}{b}} \geq 9$$

Equality holds for $a = b = c = 1$.

Comment 20.

Cauchy-Schwarz inequality *is often used to prove inequalities with square roots of the form* $\sum_{cyc} \sqrt{A} \leq k$. *On the other hand, proving inequalities of the form* $\sum_{cyc} \sqrt{A} \geq k$ *is not as straightforward. One way to tackle this type of problems is to square both sides and then using **Cauchy-Schwarz** or **Hölder's inequalities** to deal with the remaining square roots. The following examples explain this technique in more details. (You can also check **Solution 25**)*

○ **First example:** *[AoPS]*

Let a, b, c *be positive real numbers such that* $a^2 + b^2 + c^2 = 1$. *Prove that:*

$$\sqrt{1 - \left(\frac{a+b}{2}\right)^2} + \sqrt{1 - \left(\frac{b+c}{2}\right)^2} + \sqrt{1 - \left(\frac{c+a}{2}\right)^2} \geq \sqrt{6}$$

○ **Proof:**

Squaring both sides of the inequality, we get

$$\sum_{cyc} \sqrt{1 - \left(\frac{a+b}{2}\right)^2} \cdot \sqrt{1 - \left(\frac{b+c}{2}\right)^2} \geq \frac{7 + ab + bc + ca}{4}$$

On the other hand, notice that

$$1 - \left(\frac{a+b}{2}\right)^2 = \frac{(a-b)^2}{4} + \frac{c^2 + 1}{4}$$

*According to **Cauchy-Schwarz inequality**, we obtain*

$$\sum_{cyc} \sqrt{1 - \left(\frac{a+b}{2}\right)^2} \cdot \sqrt{1 - \left(\frac{b+c}{2}\right)^2} \geq \frac{(a-b)(c-b)}{4} + \frac{\sqrt{(c^2+1)(a^2+1)}}{2}$$

$$\geq \frac{b^2 + ac - ab - bc}{4} + \frac{ac + 1}{2}$$

$$= \frac{b^2 + 3ac - ab - bc + 2}{4}$$

Similarly, we obtain two other such inequalities. Summing up all of them, we get

$$\sum_{cyc} \sqrt{1 - \left(\frac{a+b}{2}\right)^2} \cdot \sqrt{1 - \left(\frac{b+c}{2}\right)^2} \geq \sum_{cyc} \frac{b^2 + 3ac - ab - bc + 2}{4}$$

$$= \frac{7 + ab + bc + ca}{4}$$

The proof is complete. Equality holds for $a = b = c = \frac{\sqrt{3}}{3}$.

○ **Second example:** [AoPS]

Let a, b, c be positive real numbers. Prove that:

$$\sum_{cyc} \sqrt{ab(a+b)^3} \geq (a+b+c)\sqrt{(a+b)(b+c)(c+a)}$$

○ **Proof:**

After squaring both sides, we need to prove that

$$\sum_{cyc} \sqrt{a(a+b)^3(a+c)^3} \geq \sqrt{abc} \sum_{cyc}(3a^2 + 5ab)$$

According to **Hölder's inequality**, we obtain

$$\left(\sum_{cyc} \sqrt{a(a+b)^3(a+c)^3}\right)^2 \left(\sum_{cyc} \frac{1}{a}\right) \geq \left(\sum_{cyc}(a^2 + 3ab)\right)^3$$

Therefore, it remains to prove that

$$\left(\sum_{cyc}(a^2 + 3ab)\right)^3 \geq (ab + bc + ca)\left(\sum_{cyc}(3a^2 + 5ab)\right)^2$$

Let $a^2 + b^2 + c^2 = t(ab + bc + ca)$. Hence, we need to prove that

$$(t+3)^3 \geq (3t+5)^2$$

or equivalently

$$(t-1)^2(t+2) \geq 0$$

which is obviously true. Equality holds for $a = b = c$.

∎

5.2   4-variable inequalities

5.2.1   Symmetric inequalities

Solution 78.

We need to prove that

$$\sum_{cyc}(a^2 - 3a + 2) \geq 0$$

Or

$$\sum_{cyc}\left((a-1)(a-2) + \frac{1}{2}\left(a - \frac{1}{a}\right)\right) \geq 0$$

which can be written as

$$\sum_{cyc} \frac{(a-1)^2(2a-1)}{a} \geq 0$$

Consequently, the inequality is obviously true if $\min\{a, b, c, d\} \geq \frac{1}{2}$. *Now, let* $a \leq \frac{1}{2}$ *and consider the following function*

$$f(x) = x^2 - 3x + 2$$

Therefore

$$\sum_{cyc} f(a) \geq f\left(\frac{1}{2}\right) + 3f\left(\frac{3}{2}\right) = 0$$

The proof is complete.

∎

Solution 79.

We will use **Mixing Variables method.** *Let*

$$f(a, b, c, d) = 18 + 2abcd - 5(a + b + c + d)$$

We will prove

$$f\left(a, b, \sqrt{\frac{c^2 + d^2}{2}}, \sqrt{\frac{c^2 + d^2}{2}}\right) \leq f(a, b, c, d)$$

which is equivalent to prove

$$ab(2cd - c^2 - d^2) - 5(c + d - \sqrt{2(c^2 + d^2)}) \geq 0$$

Or

$$\frac{5(c - d)^2}{c + d + \sqrt{2(c^2 + d^2)}} \geq ab(c - d)^2$$

Applying **AM-GM inequality,** *we deduce*

$$4 = a^2 + b^2 + (c^2 + d^2) \geq 3 \cdot \left(a^2 b^2 (c^2 + d^2)\right)^{\frac{1}{3}}$$

hence

$$ab \cdot \sqrt{\frac{c^2 + d^2}{2}} \leq \sqrt{\frac{32}{27}}$$

Therefore

$$\left(c + d + \sqrt{2(c^2 + d^2)}\right) ab \leq 4ab \cdot \sqrt{\frac{c^2 + d^2}{2}} \leq 4 \cdot \sqrt{\frac{32}{27}} < 5$$

as desired.

Finally, we can conclude that

$$f(a, b, c, d) \geq f(1, 1, 1, 1) = 0$$

Equality holds for $a = b = c = 1$.

∎

Solution 80.

Let

$$f(a,b,c,d) = \sum_{cyc} a^4 - 4abcd - 32\prod_{cyc}(a-1)$$

Consider

$$f\left(\frac{a+b}{2}, \frac{a+b}{2}, c, d\right) = \frac{(a+b)^4}{8} + c^4 + d^4 - (a+b)^2cd - 32\left(\frac{a+b}{2}-1\right)^2(c-1)(d-1)$$

Thus

$$L = f(a,b,c,d) - f\left(\frac{a+b}{2}, \frac{a+b}{2}, c, d\right)$$

$$= \frac{4(a-b)^2(a^2+ab+b^2)+3(a^2-b^2)^2}{8} + (a-b)^2cd + 8(c-1)(d-1)(a-b)^2$$

$$= \frac{(a-b)^2}{8}\left[4(a^2+ab+b^2)+3(a+b)^2+8cd+64(c-1)(d-1)\right]$$

The last line is non-negative if $(c-1)(d-1) \geq 0$.

WLOG, we can assume $a \geq b \geq 1 \geq c \geq d$ (otherwise, it is easy).

Let $x = \frac{a+b}{2}$ and $y = \frac{c+d}{2}$. Therefore

$$f(a,b,c,d) \geq f(x,x,y,y)$$

where $x + y = 2$.

Since

$$f(x,x,y,y) = 2x^4 + 2y^4 - 4x^2y^2 - 32(x-1)^2(y-1)^2$$

$$= 2(x-y)^2(x+y)^2 - 2(x-y)^4$$

$$= 2(x-y)^2((x+y)^2 - (x-y)^2)$$

$$= 8xy(x-y)^2$$

$$\geq 0$$

as desired. The proof is complete.

∎

Solution 81.

We will give four solutions to this inequality.

● *First Solution:*

*We will use **Cauchy Reverse technique** to prove this inequality*

$$\frac{1}{1+x^2} = \frac{1+x^2-x^2}{1+x^2}$$

$$= 1 - \frac{x^2}{1 + x^2}$$

$$\geq 1 - \frac{x^2}{2x}$$

$$\geq 1 - \frac{x}{2}$$

Therefore

$$\sum_{cyc} \frac{1}{1 + a^2} \geq \sum_{cyc} \left(1 - \frac{a}{2}\right) = 2$$

as desired. Equality holds for $a = b = c = d = 1$.

• **Second Solution:**

Let

$$f(u) = \frac{1}{1 + u^2}$$

Then, the inequality is equivalent to

$$f(a) + f(b) + f(c) + f(d) \geq 4f(1)$$

On the other hand

$$f''(u) = \frac{6x^2 - 2}{(x^2 + 1)^3}$$

*f is convex on $[1, 4]$. According to **Vasc's HCF theorem**, it is enough to check*

$$f(y) + 3f(x) \geq 4f(1)$$

with $y + 3x = 4$ and $x \in [0, \frac{4}{3}]$.

After some simplifications, we get

$$\frac{6(x - 1)^2(-3x^2 + 2x + 3)}{(x^2 + 1)(9x^2 - 24x + 17)} \geq 0$$

which is true for $x \in [0, \frac{4}{3}]$.

• **Third Solution:**

WLOG, we can assume that $a \leq b \leq c \leq d$. Denote

$$L(a, b, c, d) = \sum_{cyc} \frac{1}{a^2 + 1} - 2$$

We have

$$L(a, b, c, d) - L(a, \frac{b+d}{2}, c, \frac{b+d}{2}) = \frac{(b - d)^2(b^2 + 4bd + d^2 - 2)}{(b^2 + 1)(d^2 + 1)(b^2 + 2bd + d^2 + 4)}$$

On the other hand

$$b + d \geq \frac{a + b + c + d}{2} \geq 2$$

Therefore

$$L(a, b, c, d) \geq L\left(a, \frac{b+d}{2}, c, \frac{b+d}{2}\right)$$

According to **Strong Mixing Variables theorem**, *we only need to check the case* $b = c = d$. *In this case, we get* $a = 4 - 3b$ *and* $b \in [0, \frac{4}{3}]$. *We need to check*

$$L(4 - 3b, b, b, b) = \frac{6(b-1)^2(3 + 2b - 3b^2)}{(b^2 + 1)(9b^2 - 24b + 17)} \geq 0$$

which is true for $b \in [0, \frac{4}{3}]$.

This ends the proof. Equality holds for $a = b = c = d = 1$.

- **Fourth Solution:**

Using **Tangent Line method**, *we have for* $x \geq 0$

$$\frac{1}{x^2 + 1} - \frac{2 - x}{2} = \frac{x(x-1)^2}{2(x^2 + 1)} \geq 0$$

Therefore

$$\sum_{cyc} \frac{1}{a^2 + 1} \geq \sum_{cyc} \frac{2 - a}{2}$$
$$= \frac{8 - \sum_{cyc} a}{2}$$
$$= 2$$

Equality holds for $a = b = c = d = 1$.

Comment 21.

The **Tangent Line method** *is usually employed when we can rewrite the original inequality as the sum of one-variable expressions. The technique requires finding the equation of the tangent line at the equality case and then using it to bound each term of the inequality. We will show in the following examples some applications of this technique.*

○ **First Example:** *[AoPS]*

Let a, b, c, d *be positive real numbers such that* $a + b + c + d = 1$. *Prove that:*

$$6(a^3 + b^3 + c^3 + d^3) \geq a^2 + b^2 + c^2 + d^2 + \frac{1}{8}$$

○ **Proof:**

Let $f(x) = 6x^3 - x^2$. *We can rewrite the original inequality as*

$$f(a) + f(b) + f(c) + f(d) \geq \frac{1}{8}$$

On the other hand, the equation of the tangent line at $x = \frac{1}{4}$ is

$$y = f\left(\frac{1}{4}\right) + f'\left(\frac{1}{4}\right)\left(x - \frac{1}{4}\right) = \frac{5x - 1}{8}$$

Therefore, it is sufficient to prove

$$f(x) \geq \frac{5x - 1}{8}$$

which can be done as follows

$$f(x) - \frac{5x - 1}{8} = \frac{1}{8}(3x + 1)\left(x - \frac{1}{4}\right)^2 \geq 0$$

Consequently

$$\sum_{cyc} f(a) \geq \sum_{cyc} \frac{5a - 1}{8} = \frac{1}{8}$$

The proof is complete. Equality holds for $a = b = c = d = \frac{1}{4}$.

○ **Second Example:** [USAMO 2003]

Let a, b, c be positive real numbers. Prove that:

$$\frac{(2a + b + c)^2}{2a^2 + (b + c)^2} + \frac{(2b + c + a)^2}{2b^2 + (c + a)^2} + \frac{(2c + a + b)^2}{2c^2 + (a + b)^2} \leq 8$$

○ **Proof:**

WLOG, we can assume that $a + b + c = 3$. Let

$$f(x) = \frac{(x + 3)^2}{2x^2 + (3 - x)^2}$$

Therefore, we can rewrite the original inequality as follows

$$f(a) + f(b) + f(c) \leq 8$$

On the other hand, the equation of the tangent line at $x = 1$ is

$$y = f(1) + f'(1)(x - 1) = \frac{4(x + 1)}{3}$$

Now, it is enough to prove

$$f(x) \leq \frac{4(x + 1)}{3}$$

which can be done as follows

$$\frac{4(x + 1)}{3} - f(x) = \frac{(x - 1)^2(4x + 3)}{3(x^2 - 2x + 3)} \geq 0$$

Consequently

$$\sum_{cyc} f(a) \leq \sum_{cyc} \frac{4(a + 1)}{3} = 8$$

The proof is complete. Equality holds for $a = b = c = 1$.

◦ ***Third example:*** *[AoPS]*

Let a, b, c be positive real numbers such that $a + b + c = 1$. Prove that:

$$\frac{1}{a+b} + \frac{1}{b+c} + \frac{1}{c+a} + 3(ab + bc + ca) \geq \frac{11}{2}$$

◦ ***Proof:***

Let $f(x) = \frac{1}{1-x} + \frac{3x(1-x)}{2}$. Therefore, we can rewrite the inequality as

$$f(a) + f(b) + f(c) \geq \frac{11}{2}$$

On the other hand, the equation of the tangent line at $x = \frac{1}{3}$ is

$$y = f\left(\frac{1}{3}\right) + f'\left(\frac{1}{3}\right)\left(x - \frac{1}{3}\right) = \frac{11(3x+1)}{12}$$

Therefore, it is sufficient to prove

$$f(x) \geq \frac{11(3x+1)}{12}$$

which can be done as follows

$$f(x) - \frac{11(3x+1)}{12} = \frac{(2x+1)(x-\frac{1}{3})^2}{12(1-x)} \geq 0$$

Finally

$$\sum_{cyc} f(a) \geq \sum_{cyc} \frac{11(3a+1)}{12} = \frac{11}{2}$$

as desired. The equality holds for $a = b = c = \frac{1}{3}$.

■

Solution 82.

*We will use **Lagrange Multipliers**. Let*

$$g(a, b, c, d) = a + b + c + d + abcd - 5$$

$$f(a, b, c, d) = 3(a^2 + b^2 + c^2 + d^2) + 40 - 13(a + b + c + d)$$

$$L(a, b, c, d) = f(a, b, c, d) + \lambda g(a, b, c, d)$$

and

$$C = \{(a, b, c, d)|a \geq 0, b \geq 0, c \geq 0, d \geq 0, g(a, b, c, d) = 0\}$$

Note that $\nabla g = (1 + bcd, 1 + acd, 1 + abd, 1 + abc) \neq 0$ at all points, and moreover that f and g are continuous with continuous partial derivatives. On the other hand, C is compact and L is a continuous function, which says that L gets on C the minimal value.

Let this minimum occurs in the point (a, b, c, d). We have the following two cases:

(i) **Case 1:** $abcd = 0$.

WLOG, let $d = 0$. Thus, our condition gives $a + b + c = 5$ and we need to prove

$$3(a^2 + b^2 + c^2) + \frac{8(a + b + c)^2}{5} \geq \frac{13(a + b + c)^2}{5}$$

Or

$$3(a^2 + b^2 + c^2) \geq (a + b + c)^2,$$

which is true by **Cauchy-Schwarz inequality**.

(ii) **Case 2:** $abcd > 0$.

Finding the critical points, we obtain

$$\frac{\partial L}{\partial a} = \frac{\partial L}{\partial b} = \frac{\partial L}{\partial c} = \frac{\partial L}{\partial d} = 0,$$

which gives

$$\begin{cases} 6a - 13 + \lambda(1 + bcd) = 0 \\ 6b - 13 + \lambda(1 + acd) = 0 \\ 6c - 13 + \lambda(1 + abd) = 0 \\ 6d - 13 + \lambda(1 + abc) = 0 \end{cases}$$

Let $k = abcd$. Therefore, a, b, c, d are roots of

$$6x^2 - 13x + \lambda(x + k) = 0$$

which is a quadratic, thus at most has two different roots.

Consequently, we only need to check two cases:

a) $b = c = a$. Thus, we get

$$d = \frac{5 - 3a}{1 + a^3}$$

where $0 < a < \frac{5}{3}$ and we need to prove that

$$(a - 1)^2(5 - 3a)(10 + 8a + 6a^2 + 7a^3 + 2a^4 - 3a^5) \geq 0,$$

which is obvious.

b) $b = a$, $c = d$. Thus, the condition gives

$$2(a + d) + a^2d^2 = 5$$

and we need to prove that

$$3(a^2 + d^2) + 20 \geq 13(a + d).$$

Now, let $a + d = 2u$ and $ad = v^2$. Thus, $v^2 \leq u^2$, $u = \frac{5 - v^4}{4}$, which gives $v^2 \leq \sqrt{5}$ and from here $v^2 \leq 1$.

Finally, we only need to prove that

$$3\left(\frac{(5-v^4)^2}{8} - v^2\right) + 10 \geq \frac{13(5-v^4)}{4}$$

which is equivalent to

$$(1 - v^2)(25 + v^2 - 3v^4 - 3v^6) \geq 0$$

which is obvious. The proof is complete.

∎

Solution 83.

*For $d = 0$, the inequality is true by **AM-GM inequality**. If $abcd \neq 0$, we can rewrite our inequality in the following form*

$$27abcd\left(\frac{1}{a} + \frac{1}{b} + \frac{1}{c} + \frac{1}{d}\right) + 17abcd \leq 125$$

*According to **Vasc's EV theorem**, it is enough to prove our inequality for $b = c = a$ and $d = \frac{5-3a}{1+a^3}$, where $0 < a < \frac{5}{3}$, which gives*

$$(a-1)^2(5-3a)(9a^3 + 33a^2 + 65a + 25) \geq 0$$

as desired.

∎

Solution 84.

*Applying the **Buffalo Way technique**: let $a = \min\{a, b, c, d\}$, $b = a + u$, $c = a + v$ and $d = a + w$. Therefore*

$$I = \sum_{cyc}(a-b)(a-c)(a-d) + \frac{1}{27}(a+b+c+d)^3$$

$$= -uvw + \sum_{cyc}u(u-v)(u-w) + \frac{1}{27}(4a+u+v+w)^3$$

$$\geq \sum_{cyc}u(u-v)(u-w) + \frac{1}{27}(u+v+w)^3 - uvw$$

$$\geq 0$$

*where the last inequality is true by **Schur's** and **AM-GM inequalities**.*

∎

Solution 85.

We will give two solutions to this inequality.

● *First Solution:*

WLOG, assume $a \geq b \geq c \geq d$. Denote

$$L(a, b, c, d) = a^2 + b^2 + c^2 + d^2 + abcd + 1 - (ab + bc + cd + da + ac + bd)$$

We have

$$L(a, b, c, d) - L(\sqrt{ac}, b, \sqrt{ac}, d) = (\sqrt{a} - \sqrt{c})^2(a + c - b - d + 2\sqrt{ac}) \geq 0$$

*According to **Strong Mixing Variables method**, we only need to check $a = b = c$. In this case, we get*

$$L(a, a, a, d) = a^3 d + d^2 + 1 - 3ad \geq 0$$

*which obviously true by **AM-GM inequality**.*

This ends the proof. Equality holds for $(1, 1, 1, 1)$.

● *Second Solution:*

*This inequality is a straightforward application of **Turkevich's inequality***

$$a^2 + b^2 + c^2 + d^2 + abcd + 1 \geq a^2 + b^2 + c^2 + d^2 + 2\sqrt{abcd}$$
$$\geq ab + bc + cd + da + ac + bd$$

as desired. Equality holds for $a = b = c = d = 1$.

Comment 22.

*We will give three proofs of **Turkevich's inequality***

○ *First proof:*

*We will use **Karamata's inequality** to prove this inequality. Indeed, let $a = e^x$, $b = e^y$, $c = e^z$ and $d = e^t$. We can rewrite **Turkevich's inequality** as*

$$\sum_{cyc} e^{4x} + 2e^{x+y+z+t} \geq \sum_{sym} e^{2x+2y}$$

Or $f(x) = e^x$ is convex, it is sufficient to prove that (A) majorizes (B) with

$$(A) = (4x, 4y, 4z, 4t, x + y + z + t, x + y + z + t)$$

$$(B) = (2x + 2y, 2y + 2z, 2z + 2t, 2t + 2x, 2x + 2z, 2y + 2t)$$

which is not hard to check.

○ *Second proof:*

WLOG, we can assume $a \geq b \geq c \geq d$. Thus, **Turkevich's inequality** *can be rewritten as*

$$(ab + cd - c^2 - d^2)^2 + 2cd(c - d)^2 + (a - b)^2((a + b)^2 - (c^2 + d^2)) \geq 0$$

which is obviously true.

○ *Third proof:*

Let $x_1 = a$, $x_2 = b$, $x_3 = c$ and $x_4 = d$. Applying **Suranyi's inequality** *for $n = 4$, we obtain*

$$3\sum_{i=1}^{4} x_i^4 + 4\prod_{i=1}^{4} x_i \geq \left(\sum_{i=1}^{4} x_i\right)\left(\sum_{i=1}^{4} x_i^3\right)$$

which can be rewritten as

$$2\left(\sum_{i=1}^{4} x_i^4 + 2\prod_{i=1}^{4} x_i\right) \geq \sum_{1 \leq i < j \leq 4} x_i x_j (x_i^2 + x_j^2)$$

On the other hand, we know that $x_i^2 + x_j^2 \geq 2x_i x_j$, therefore

$$\sum_{i=1}^{4} x_i^4 + 2\prod_{i=1}^{4} x_i \geq \sum_{1 \leq i < j \leq 4} x_i^2 x_j^2$$

which is **Turkevich's inequality**.

■

Solution 86.

First of all, we have

$$\sum_{cyc} \frac{1}{a+1} - 2 = \frac{a+b+c+d - abc - bcd - cda - dab + 2(1 - abcd)}{(1+a)(1+b)(1+c)(1+d)}$$

On the other hand, we know that

$$abcd \leq 1$$

Therefore, it is sufficient to prove

$$a + b + c + d \geq abc + bcd + cda + dab$$

Using **Problem 83**, *we get*

$$27(abc + bcd + acd + abd) + 17abcd \leq 125$$

Therefore

$$a + b + c + d - (abc + bcd + cda + dab) \geq a + b + c + d - \frac{125 - 17abcd}{27}$$

$$= \frac{27(a+b+c+d)+17abcd-125}{27}$$

$$= \frac{10(a+b+c+d)+85-125}{27}$$

$$= \frac{10(a+b+c+d-4)}{27}$$

$$\geq 0$$

The problem is completely solved. Equality holds for $a = b = c = d = 1$.

∎

Solution 87.

Rewriting the original inequality in the homogeneous form, we get

$$(a^3 + b^3 + c^3 + d^3 + abc + bcd + cda + dab)^2 \leq (a^2 + b^2 + c^2 + d^2)^3$$

Let

$$S = a^3 + b^3 + c^3 + d^3 + abc + bcd + cda + dab$$

Notice that

$$a^3 + b^3 + c^3 + d^3 + abc + bcd + cda + dab = \sum_{cyc} a(a^2 + bc)$$

*By **Cauchy-Schwarz inequality**, we find that*

$$S^2 \leq (a^2 + b^2 + c^2 + d^2)\left[(a^2 + bc)^2 + (b^2 + cd)^2 + (c^2 + da)^2 + (d^2 + ab)^2\right]$$

It's sufficient to prove that

$$(a^2 + bc)^2 + (b^2 + cd)^2 + (c^2 + da)^2 + (d^2 + ab)^2 \leq (a^2 + b^2 + c^2 + d^2)^2$$

which is equivalent to

$$2(a^2bc + b^2cd + c^2da + d^2ab) \leq a^2b^2 + c^2d^2 + a^2d^2 + b^2c^2 + 2(a^2c^2 + b^2d^2)$$

This last inequality is true as it can be transformed into

$$(ab - ac)^2 + (ac - cd)^2 + (bc - bd)^2 + (ad - bd)^2 \geq 0$$

This ends the proof.

∎

Solution 88.

Let $a - \frac{1}{a} = x$, $b - \frac{1}{b} = y$, $c - \frac{1}{c} = z$ and $d - \frac{1}{d} = t$. Therefore

$$x + y + z + t = 0$$

We need to prove that

$$\sum_{cyc} f(x) \geq 0$$

where

$$f(x) = \ln\left(x + 2 + \sqrt{x^2 + 4}\right) - 2\ln 2$$

On the other hand, we have

$$f''(x) = -\frac{\left(x + \sqrt{x^2+4}\right)\left(x^2 + 2x + (x-2)\sqrt{x^2+4}\right)}{\sqrt{(x^2+4)^3}\left(x + 2 + \sqrt{x^2+4}\right)^2}$$

which gives that $f''(x) = 0$ for a unique number $x = 0.86...$

*According to **Vasc's HCF theorem**, it is enough to prove our inequality for $y = z = x$ and $t = -3x$, which is $b = c = a$ and $d = \frac{3 - 3a^2 + \sqrt{9 - 14a^2 + 9a^4}}{2a}$.*

In this case, we get

$$(1+a)^3(3 + 2a - 3a^2 + \sqrt{9a^4 - 14a^2 + 9}) \geq 32a$$

Or

$$(1+a)^3\sqrt{9a^4 - 14a^2 + 9} \geq 3a^5 + 7a^4 - 12a^2 + 21a - 3$$

Since

$$(1+a)^3\sqrt{9a^4 - 14a^2 + 9} > 0$$

It is enough to prove that

$$(1+a)^6(9a^4 - 14a^2 + 9) \geq (3a^5 + 7a^4 - 12a^2 + 21a - 3)^2$$

which is

$$4a(a-1)^2(3a^6 + 24a^5 + 87a^4 + 144a^3 + 89a^2 - 8a + 45) \geq 0$$

This ends the proof. Equality holds for $a = b = c = d = 1$.

∎

Solution 89.

Let $a+b+c+d = 4u$, $ab+bc+cd+da+ac+bd = 6v^2$, $abc+abd+acd+bcd = 4w^3$ and $abcd = t^4$. Therefore, the condition gives

$$ut^4 = w^3$$

We need to prove that

$$16u^2 - 12v^2 + 2t^4 + 6 \geq 12u$$

Or

$$\frac{(8u^2 - 6v^2)ut^4}{w^3} + t^4 + \frac{3u^2 t^8}{w^6} \geq 6u\left(\sqrt{\frac{ut^4}{w^3}}\right)^3$$

Or

$$3u^2 t^4 - 6u\sqrt{u^3 w^3}\, t^2 + (8u^3 - 6uv^2 + w^3)w^3 \geq 0$$

which is a quadratic on t^2. Therefore, it is enough to prove that

$$\Delta = 9u^5 w^3 - 3u^2(8u^3 - 6uv^2 + w^3)w^3 \leq 0$$

which is equivalent to

$$5u^3 - 6uv^2 + w^3 \geq 0$$

According to **Rolle's theorem,** *there exist non-negative real numbers x, y and z for which*

$$\begin{cases} 3u = x + y + z \\ 3v^2 = xy + yz + zx \\ w^3 = xyz \end{cases}$$

Thus, by **Schur's inequality,** *we obtain*

$$w^3 \geq 4uv^2 - 3u^3$$

Finally

$$5u^3 - 6uv^2 + w^3 \geq 5u^3 - 6uv^2 + 4uv^2 - 3u^3$$
$$= 2u(u^2 - v^2)$$
$$\geq 0$$

The proof is complete.

Comment 23.

Rolle's theorem *is often used to reduce the number of variables, especially when dealing with 4-variable inequalities. We will show in the following examples how to apply this technique to prove some difficult problems.*

○ **First example:** *[AoPS]*

Let a, b, c, d be non-negative real numbers. Prove that:

$$\frac{2}{3}(ab + bc + cd + da + ac + bd) \geq \sqrt{(abc + bcd + cda + dab)(a + b + c + d)}$$

○ **Proof:**

Let $a + b + c + d = 4u$, $ab + bc + cd + da + ac + bd = 6v^2$ and $abc + bcd + cda + dab = 4w^3$. We only need to prove

$$v^2 \geq \sqrt{uw^3}$$

According to **Rolle's Theorem,** *there exist x, y and z such that $x + y + z = 3u$, $xy + yz + zx = 3v^2$ and $xyz = w^3$. Replacing in the above inequality, we need to prove*

$$(xy + yz + zx)^2 \geq 3xyz(x + y + z)$$

which is obvious. Equality holds for $a = b = c = d$.

○ **Second example:** *[AoPS]*

Let $a, b, c, d \geq 0$ such taht $a + b + c + d = 1$. Prove that:

$$(1 - a)(1 - b)(1 - c)(1 - d) \leq abcd + \frac{5}{16}$$

○ ***Proof:***

Let $a + b + c + d = 4u$, $ab + bc + cd + da + ac + bd = 6v^2$ and $abc + bcd + cda + abd = 4w^3$. Thus, we need to prove that

$$5u^3 + w^3 \geq 6uv^2$$

According to **Rolle's theorem**, there exist positive numbers x, y and z for which $3u = x + y + z$, $3v^2 = xy + yz + zx$ and $w^3 = xyz$. Using **Schur's inequality**, we obtain

$$w^3 \geq 4uv^2 - 3u^3$$

Therefore

$$5u^3 + w^3 \geq 5u^3 + 4uv^2 - 3u^3$$
$$= 2u^3 + 4uv^2$$
$$\geq 6uv^2$$

as desired. Equality holds for $a = b = c = d$.

○ ***Third example:*** *[Vietnam MO 1996]*

Let a, b, c, d be positive real numbers such that:

$$2(ab + bc + cd + da + ac + bd) + abc + bcd + cda + dab = 16$$

Prove that:

$$3(a + b + c + d) \geq 2(ab + bc + cd + da + ac + bd)$$

○ ***Proof:***

Let $a + b + c + d = 4u$, $ab + bc + cd + da + ac + bd = 6v^2$ and $abc + bcd + cda + abd = 4w^3$. Thus, we need to prove that

$$u \geq v^2$$

According to **Rolle's theorem**, there exist positive numbers x, y and z for which $3u = x + y + z$, $3v^2 = xy + yz + zx$ and $w^3 = xyz$. Therefore, we have $xy + yz + zx + xyz = 4$ and it remains to prove that

$$x + y + z \geq xy + yz + zx$$

Because $x, y, z \geq 0$ such that $xy + yz + zx + xyz = 4$ so there exist $m, n, p \geq 0$ satisfying

$$\begin{cases} x = \frac{2m}{n+p} \\ y = \frac{2n}{p+m} \\ z = \frac{2p}{m+n} \end{cases}$$

Therefore, it is sufficient to prove

$$\sum_{cyc} \frac{2m}{n+p} \geq \sum_{cyc} \frac{4mn}{(n+p)(p+m)}$$

that we can rewrite as

$$\sum_{cyc} m(m+n)(m+p) \geq \sum_{cyc} 2mn(m+n)$$

or equivalenty

$$\sum_{cyc} m^3 + 3mnp \geq \sum_{cyc} mn(m+p)$$

*which is true by **Schur's inequality**. Equality holds for $a = b = c = d = 1$.*

■

Solution 90.

We will give two solutions to this problem.

• First Solution:

It's equivalent to $\sum_{cyc} f(a) \geq 0$ where

$$f(x) = \frac{a^2}{4-a} - \frac{a^2}{2} + \frac{1}{6}$$

on $[0, 4)$.

*Since f is strictly convex on $[1, 4)$, by **Vasc's HCF theorem**, it suffices to prove that for $\forall a \in \left(0, \frac{4}{3}\right]$, we have*

$$3f(a) + f(4 - 3a) \geq 0$$

that can be transformed into

$$(a-1)^2(4-3a)(8-3a) \geq 0$$

which is obviously true $\forall a \in \left(0, \frac{4}{3}\right]$.

Equality holds for $(1, 1, 1, 1)$, $\left(\frac{4}{3}, \frac{4}{3}, \frac{4}{3}, 0\right)$ and permutations.

• Second Solution:

Let

$$L(a, b, c, d) = \sum_{cyc} \left(\frac{a^2}{4-a} - \frac{a^2}{2} + \frac{1}{6} \right)$$

We aim to prove

$$L(a, b, c, d) \geq 0$$

WLOG, we can assume $a \geq b \geq c \geq d$. Therefore

$$a + c \geq \frac{a+b+c+d}{2} = 2$$

Let $t = \frac{a+c}{2}$. We have

$$L(a, b, c, d) - L(t, b, t, d) = \frac{(a-c)^2}{4}\left(\frac{64}{(4-a)(4-c)(8-a-c)} - 1\right)$$

On the other hand

$$(4-a)(4-c)(8-a-c) \leq \frac{(8-a-c)^3}{4} \leq \frac{6^3}{4} = 54 < 64$$

We conclude that

$$L(a, b, c, d) \geq L(t, b, t, d)$$

According to **Strong Mixing Variables method**, we only need to check the case $a = b = c$. In this case, we get $d = 4 - 3a$, and we need to check

$$L(a, a, a, 4 - 3a) = \frac{2(a-1)^2(8-3a)(4-3a)}{3a(4-a)} \geq 0$$

which is obviously true for $a \in [0, \frac{4}{3}]$.

as desired. Equality holds for $(1, 1, 1, 1)$, $\left(\frac{4}{3}, \frac{4}{3}, \frac{4}{3}, 0\right)$ and permutations.

∎

Solution 91.

We will give three solutions to this problem.

• First Solution

The equality is equivalent to

$$3(a^2 + b^2 + c^2 + d^2) + 4abcd \geq 16$$

Fix the value of c, d, so $a + b$ is also fixed. Now consider

$$f(ab) = ab(4cd - 6) + 3(a+b)^2 + 3(c+d)^2 - 6cd$$

above $\left[0, \frac{(a+b)^2}{4}\right]$.

Since f is linear, f gets a minimal value when either $ab = 0$ or $ab = \frac{(a+b)^2}{4}$, that is $a = b$. Similarly, we get $cd = 0$ or $cd = \frac{(c+d)^2}{4}$. Therefore, we have two cases:

(i) **The first case:** $abcd = 0$.

WLOG, assume that $d = 0$. Therefore

$$LHS \geq 3(a^2 + b^2 + c^2) \geq (a + b + c)^2 = 16$$

(ii) **The second case:** $a = b$ and $c = d$.

Therefore $a + c = 2$ and

$$LHS = 6(a^2 + c^2) + 4a^2c^2 = 5(a + c)^2 + (a - c)^2 + 4(ac - 1)^2 - 4 \geq 16$$

The proof is complete.

- ## *Second Solution*

Let

$$L(a, b, c, d) = 3(a^2 + b^2 + c^2 + d^2) + 4abcd$$

WLOG, we can assume $a \geq b \geq c \geq d$. Let $a + b = x$ and $c + d = y$.

There are two cases:

(i) **The first case:** $ab \geq \frac{3}{2}$.

$$L(a, b, c, d) = 3(a^2 + b^2 + c^2 + d^2) + 4abcd$$
$$\geq \frac{3x^2}{2} + 3y^2$$
$$= \frac{(x - 2y)^2}{2} + (x + y)^2$$
$$\geq 16$$

(ii) **The second case:** $ab \leq \frac{3}{2}$.

$$L(a, b, c, d) - 16 = 3(a^2 + b^2 + c^2 + d^2) + 4abcd - 16$$
$$= \frac{1}{2}(c - d)^2(3 - 2ab) + \frac{1}{4}(a - b)^2(6 - y^2) + \frac{1}{4}(x - 2)^2(x^2 - 4x + 8)$$
$$\geq 0$$

as desired.

The proof is complete.

- ## *Third Solution:*

*We will use **Lagrange Multipliers**. Let*

$$f(a.b, c, d) = 3(a^2 + b^2 + c^2 + d^2) + 4abcd - 16$$

$$g(a, b, c, d) = a + b + c + d - 4$$

$$L(a, b, c, d) = 3(a^2 + b^2 + c^2 + d^2) + 4abcd - 16 + \lambda(a + b + c + d - 4)$$

and

$$C = \{(a, b, c, d) | a \geq 0, b \geq 0, c \geq 0, d \geq 0, a + b + c + d = 4\}$$

Note that $\nabla g = (1, 1, 1, 1) \neq 0$ at all points, and moreover that f and g are continuous with continuous partial derivatives. On the other hand, C is compact and L is a continuous function, which says that L gets on C the minimal value.

Let this minimum occurs in the point (a, b, c, d). We have the following two cases:

(i) **The first case:** $abcd = 0$.

Let $d = 0$. Therefore, our condition gives $a + b + c = 4$ and we need to prove that

$$3(a^2 + b^2 + c^2) \geq (a + b + c)^2,$$

which is **Cauchy-Schwarz inequality**.

(ii) **The second case:** $abcd > 0$.

In this case, we get

$$\frac{\partial L}{\partial a} = \frac{\partial L}{\partial b} = \frac{\partial L}{\partial c} = \frac{\partial L}{\partial d} = 0$$

which gives

$$\begin{cases} \frac{\partial L}{\partial a} = 6a + 4bcd + \lambda = 0 \\ \frac{\partial L}{\partial b} = 6b + 4acd + \lambda = 0 \\ \frac{\partial L}{\partial c} = 6c + 4abd + \lambda = 0 \\ \frac{\partial L}{\partial d} = 6d + 4abc + \lambda = 0 \end{cases}$$

If we note $k = abcd$, we can see that a, b, c, d are all roots of

$$6x^2 + \lambda x + 4k = 0$$

Therefore, we only need to check 2 cases: (x,x,y,y) or (x,x,x,y).

<u>**case 1:**</u> $(a, b, c, d) = (x, x, y, y)$

In this case, $y = 2 - x$ and we get

$$4(x - 1)^2((x - 1)^2 + 1) \geq 0$$

which is true.

<u>**case 2:**</u> $(a, b, c, d) = (x, x, x, y)$

In this case, $y = 4 - 3x$ and we get

$$4(x - 1)^2(4 - 3x)(x + 2) \geq 0$$

which is true.

Equality holds for $(1, 1, 1, 1)$, $\left(0, \frac{4}{3}, \frac{4}{3}, \frac{4}{3}\right)$ and permutations.

■

5.2.2 Non-symmetric inequalities

Solution 92.

For any real numbers x and y, we have

$$2 + x^2 + y^2 - (1 + x)(1 + y) = \frac{(2x - y - 1)^2}{4} + \frac{3(y - 1)^2}{4} \geq 0$$

Therefore

$$2 + x^2 + y^2 \geq (1+x)(1+y)$$

Finally

$$(2 + a^2 + b^2)(2 + b^2 + c^2)(2 + c^2 + d^2)(2 + d^2 + a^2) \geq (1 + |a|)^2(1 + |b|)^2(1 + |c|)^2(1 + |d|)^2$$
$$\geq (1 + a)^2(1 + b)^2(1 + c)^2(1 + d)^2$$

as desired. Equality holds for $a = b = c = d = 1$.

■

Solution 93.

*Using **weighted AM-GM**, we have that for $x \geq 0$*

$$\frac{x^i + (i-1)}{i} \geq x$$

Therefore

$$\frac{x^i}{i} \geq x - \frac{i-1}{i}$$

Applying this inequality to b,c and d, we get

$$\frac{b^2}{2} \geq b - \frac{1}{2}$$

$$\frac{c^3}{3} \geq c - \frac{2}{3}$$

$$\frac{d^4}{4} \geq d - \frac{3}{4}$$

After summing up these inequalities and using the condition $a + b + c + d = 4$ we obtain the desired inequality.

Equality holds for $a = b = c = d = 1$.

Comment 24.

*At this stage, you are probably wondering how we came up with the weights in our proof. The answer to this question is known as **the Balanced Coefficients method** which was first introduced by Pham Kim Hung in his book "Secrets in Inequalities (Volume 1)". The main idea is to introduce new variables while proving the original inequality. In the end, the equality case will determine the values of the intermediate variables. We will show in the following examples some applications of this technique.*

○ ***First example:** [IMO Shortlist 1998]*

Let a, b, c be positive real numbers such that $abc = 1$. Prove that:

$$\frac{a^3}{(1+b)(1+c)} + \frac{b^3}{(1+c)(1+a)} + \frac{c^3}{(1+a)(1+b)} \geq \frac{3}{4}$$

○ **Proof:**

We will use the **Balanced Coefficients method:** Let x and y be positive numbers. According to **AM-GM inequality**, we obtain

$$\frac{a^3}{(1+b)(1+c)} + x^3(1+b) + y^3(1+c) \geq 3xya$$

In a similar way, we get

$$\frac{b^3}{(1+c)(1+a)} + x^3(1+c) + y^3(1+a) \geq 3xyb$$

$$\frac{c^3}{(1+a)(1+b)} + x^3(1+a) + y^3(1+b) \geq 3xyc$$

Summing up all these inequalities, we get

$$\sum_{cyc} \frac{a^3}{(1+b)(1+c)} + (3+a+b+c)(x^3+y^3) \geq 3xy(a+b+c)$$

Evaluating this expression at $(1,1,1)$, we get

$$\frac{1}{8} + x^3 + y^3 = \frac{3xy}{2}$$

According to **AM-GM inequality**, we deduce that

$$x = y = \frac{1}{2}$$

In this case, we get

$$\sum_{cyc} \frac{a^3}{(1+b)(1+c)} + \frac{3+a+b+c}{4} \geq \frac{3(a+b+c)}{4}$$

or equivalently

$$\sum_{cyc} \frac{a^3}{(1+b)(1+c)} \geq \frac{a+b+c}{2} - \frac{3}{4}$$

On the other hand, we have

$$a+b+c \geq 3\sqrt[3]{abc} = 3$$

Therefore

$$\sum_{cyc} \frac{a^3}{(1+b)(1+c)} \geq \frac{3}{4}$$

The proof is complete. Equality holds for $a = b = c = 1$.

○ **Second example:** [AoPS]

Let a, b, c, d be positive real numbers. Prove that:

$$a^4b + b^4c + c^4d + d^4a \geq abcd(a+b+c+d)$$

o **Proof:**

We will use the **Balanced Coefficients method:** Let p, q, r, s be positive real numbers such that $p + q + r + s = 1$. According to **weighted AM-GM**, we obtain

$$pa^4b + qb^4c + rc^4d + sd^4a \geq a^{4p+s}b^{4q+p}c^{4r+q}d^{4s+p}$$

Now, we want $a^{4p+s}b^{4q+p}c^{4r+q}d^{4s+p} = a^2bcd$, thus, we get the following system

$$\begin{cases} 4p + s = 2 \\ 4q + p = 1 \\ 4r + s = 1 \\ 4s + p = 1 \end{cases}$$

Therefore, we get

$$(p, q, r, s) = \left(\frac{23}{51}, \frac{7}{51}, \frac{11}{51}, \frac{10}{51}\right)$$

Consequently

$$\frac{23a^4b + 7b^4c + 11c^4d + 10d^4a}{51} \geq a^2bcd$$

Similarly, we obtain

$$\frac{10a^4b + 23b^4c + 7c^4d + 11d^4a}{51} \geq ab^2cd$$

$$\frac{11a^4b + 10b^4c + 23c^4d + 7d^4a}{51} \geq abc^2d$$

$$\frac{7a^4b + 11b^4c + 10c^4d + 23d^4a}{51} \geq abcd^2$$

Summing up all these inequalities, we obtain

$$a^4b + b^4c + c^4d + d^4a \geq a^2bcd + ab^2cd + abc^2d + abcd^2$$
$$= abcd(a + b + c + d)$$

The proof is complete. Equality holds for $a = b = c = d$.

∎

Solution 94.

We have

$$\sum_{cyc}(a^4 - abcd) = \frac{1}{2}\sum_{cyc}(a - b)^4 + \sum_{cyc}(2a^3b + 2ab^3 - 3a^2b^2) - 4abcd$$

$$= \frac{1}{2}\sum_{cyc}(a - b)^4 + \sum_{cyc}(2a^3b + 2ab^3 - 4a^2b^2) + \sum_{cyc}(a^2b^2 - abcd)$$

$$\geq 2\left|\prod_{cyc}(a - b)\right|$$

$$\geq -2\prod_{cyc}(a-b)$$

The proof is complete.

∎

Solution 95.

*By **Cauchy-Schwarz inequality**, we find that*

$$\frac{1}{(a-b)^2} + \frac{1}{(b-c)^2} + \frac{1}{(c-d)^2} + \frac{1}{(d-a)^2} \geq \frac{16}{(a-b)^2+(b-c)^2+(c-d)^2+(d-a)^2}$$

It suffices to prove that

$$(a-b)^2+(b-c)^2+(c-d)^2+(d-a)^2 \leq 8$$

which is true since

$$(a-b)^2+(b-c)^2+(c-d)^2+(d-a)^2 = 2(a^2+b^2+c^2+d^2-ab-bc-cd-da)$$
$$\leq 2(a^2+b^2+b^2+c^2)$$
$$= 8$$

The proof is complete.

∎

Solution 96.

We will study two cases $d \geq 1$ and $d < 1$.

*(i) **The first case:** $d \geq 1$.*

Thus
$$a+b+c \leq 3$$

*According to **AM-GM inequality**, we get*

$$ab \leq \left(\frac{a+b}{2}\right)^2 \leq \frac{9}{4}$$

It is enough to prove that
$$a+abc \geq abcd$$

Or
$$1 \geq bc(d-1)$$

*which is true by **AM-GM inequality***

$$bc(d-1) \leq \left(\frac{b+c+d-1}{3}\right)^3 = \left(\frac{3-a}{3}\right)^3 \leq 1$$

(ii) **The second case:** $d < 1$.

It is enough to prove that
$$a + \frac{9}{4} \geq ab$$

Or
$$\frac{9}{4} \geq a(b-1)$$

which is obvious for $b < 1$. *For* $b \geq 1$, *we obtain*
$$a(b-1) \leq \left(\frac{a+b-1}{2}\right)^2 = \left(\frac{3-c-d}{2}\right)^2 \leq \frac{9}{4}$$

The proof is complete.

■

Solution 97.

WLOG, we can assume that $d \leq a, b, c$. *We have*
$$a^2b + b^2c + c^2d + d^2a = a^2b + b^2c + c^2a + abc + c^2(d-a) + a(d^2 - bc)$$

On the other hand, we know that
$$a^2b + b^2c + c^2a + abc \leq \frac{4(a+b+c)^3}{27}$$

From $d \leq a, b, c$, *we get*
$$c^2(d-a) + a(d^2 - bc) \leq 0$$

Finally
$$a^2b + b^2c + c^2d + d^2a \leq \frac{4(4-d)^3}{27} \leq \frac{4^4}{27} = \frac{256}{27}$$

Equality holds for $\left(\frac{8}{3}, \frac{4}{3}, 0, 0\right)$ *and circular permutations.*

■

Solution 98.

Notice that
$$\frac{a}{b} - \frac{a+b}{b+c} = \frac{a}{b} + 1 - \frac{a+b}{b+c} - 1$$
$$= \frac{a+b}{b} - \frac{a+b}{b+c} - 1$$
$$= \frac{(a+b)(b+c-c)}{b(b+c)} - 1$$
$$= \frac{c(a+b)}{b(b+c)} - 1$$

Summing up all the similar equations and applying **AM-GM inequality**
$$\sum_{cyc} \frac{a}{b} - \sum_{cyc} \frac{a+b}{b+c} = \sum_{cyc} \frac{c(a+b)}{b(b+c)} - 4 \geq 0$$

Solution 99.

Using ***Cauchy-Schwarz*** *and* ***AM-GM inequalities****, we get*

$$(a^2b + b^2c + c^2d + d^2a)^2 \le (a^2 + b^2 + c^2 + d^2)(a^2b^2 + b^2c^2 + c^2d^2 + d^2a^2)$$
$$= 4(a^2 + c^2)(b^2 + d^2)$$
$$\le 4\left(\frac{a^2 + b^2 + c^2 + d^2}{2}\right)^2$$
$$= 16$$

Therefore

$$-4 \le a^2b + b^2c + c^2d + d^2a \le 4$$

Equality case: Minimum is attained at $a = b = c = d = -1$, *while the Maximum is attained at* $a = b = c = d = 1$.

∎

5.3   Geometric inequalities

Solution 100.

Let S_{ABC} *be the area of the triangle* $\triangle ABC$. *We have*

$$R = \frac{abc}{4S_{ABC}}$$

Let C' *be the midpoint of* $[AB]$. *Thus*

$$S_{GBC} = \frac{2}{3}S_{C'CB} = \frac{1}{3}S_{ABC}$$

Therefore

$$R_1 = \frac{3a.GB.GC}{4S_{ABC}}$$

In a similar way, we have

$$R_2 = \frac{3b.GA.GC}{4S_{ABC}}$$

$$R_3 = \frac{3c.GA.GB}{4S_{ABC}}$$

Consequently, it is sufficient to prove

$$a.GB.GC + b.GA.GC + c.GA.GB \ge abc$$

which is true by ***Hayashi's inequality****.*

Comment 25.

Hayashi's inequality *can easily be proved using* ***Complex Numbers*** *and the* ***Triangle inequality*** *as follows:*

Let a, b, c, p be the complex coordinates of A, B, C and P respectively. We can easily check the following identity

$$\sum_{cyc}(a-b)(p-a)(p-b) = (a-b)(b-c)(c-a)$$

Therefore

$$\sum_{cyc} BC.PB.PC = \sum_{cyc}|(b-c)(p-b)(p-c)|$$

$$\geq \left|\sum_{cyc}(b-c)(p-b)(p-c)\right|$$

$$= |(a-b)(b-c)(c-a)|$$

$$= AB.BC.CA$$

The proof is complete. Note that the equality holds if and only if $P = H$, where H is the orthocenter of the triangle $\triangle ABC$.

■

Solution 101.

First of all, we have

$$\sum_{cyc}\cos\alpha = 1 + (-1 + \cos\alpha) + (\cos\beta + \cos\gamma)$$

$$= 1 - 2\sin^2\frac{\alpha}{2} + 2\cos\frac{\beta+\gamma}{2}\cos\frac{\beta-\gamma}{2}$$

$$= 1 + 2\sin\frac{\alpha}{2}\left(\cos\frac{\beta-\gamma}{2} - \cos\frac{\beta+\gamma}{2}\right)$$

$$= 1 + 4\prod_{cyc}\sin\frac{\alpha}{2}$$

$$= 1 + 4\prod_{cyc}\sqrt{\frac{(s-b)(s-c)}{bc}}$$

$$= 1 + 4\frac{(s-a)(s-b)(s-c)}{abc}$$

$$= 1 + 4\frac{sr^2}{4Rsr}$$

$$= 1 + \frac{r}{R}$$

Therefore

$$\cos\alpha + \cos\beta + \cos\gamma = 1 + \frac{r}{R}$$

○ **Left inequality:**

Applying **Euler's inequality**, we obtain

$$\cos\alpha + \cos\beta + \cos\gamma = 1 + \frac{r}{R} \leq 1 + \frac{1}{2} = \frac{3}{2}$$

(*Alternatively, we could use **Klamkin's inequality** with $n = 1$ and $x = y = z$.*)

○ **Right inequality:**

Applying **Euler's formula**, we get

$$OI = \sqrt{R(R - 2r)}$$

We need to prove

$$\frac{3}{2} - \left(1 + \frac{r}{R}\right) \le \frac{\sqrt{R(R - 2r)}}{s}$$

Or

$$s^2(R - 2r) \le 4R^3$$

Applying **Gerretsen's inequality**, we get

$$s^2 \le 4R^2 + 4Rr + 3r^2$$

It is sufficient to prove

$$(4R^2 + 4Rr + 3r^2)(R - 2r) \le 4R^3$$

Or

$$0 \le r(6r^2 + 5rR + 4R^2)$$

The inequality is proved.

Comment 26.

We will show here five different ways to prove **Euler's inequality**.

○ **First proof:**

According to **Hayashi's inequality** with $P = O$, we obtain

$$(a + b + c)R^2 \ge abc$$

Therefore

$$\begin{aligned}
R^2 &\ge \frac{abc}{a + b + c} \\
&= \frac{4R}{2s} \cdot \frac{abc}{4R} \\
&= 2R\frac{S_{ABC}}{s} \\
&= 2Rr
\end{aligned}$$

as required.

○ **Second proof:**

We proved previously that

$$(a - b)^2(b - c)^2(c - a)^2 = -4r^2[(s^2 - 2R^2 - 10Rr + r^2)^2 - 4R(R - 2r)^3]$$

Therefore

$$(s^2 - 2R^2 - 10Rr + r^2)^2 \leq 4R(R - 2r)^3$$

Finally

$$R \geq 2r$$

○ **Third proof:**

We will use the relations between inradius and exradii (r_a, r_b, r_c) to prove the inequality. We have

$$S_{ABC} = \sqrt{s(s-a)(s-b)(s-c)} = sr = \frac{abc}{4R} = r_x(s-x), x \in \{a,b,c\}$$

Therefore, we get

$$\frac{1}{r_a} + \frac{1}{r_b} + \frac{1}{r_c} = \frac{1}{r}$$

$$r_a + r_b + r_c = r + 4R$$

Now, applying **AM-GM inequality**, we get

$$(r_a + r_b + r_c)\left(\frac{1}{r_a} + \frac{1}{r_b} + \frac{1}{r_c}\right) \geq 9$$

Or

$$\frac{r + 4R}{r} \geq 9$$

Therefore

$$R \geq 2r$$

○ **Fourth proof:**

We use the following formulae for the area of a triangle

$$S = \sqrt{s(s-a)(s-b)(s-c)} = rs = \frac{abc}{4R}$$

where S is the area, r is the inradius, R is the circumradius and s is the semiperimeter. Hence

$$R \geq 2r \iff \frac{abc}{4S} \geq \frac{2S}{s} \iff \frac{abc}{8} \geq (s-a)(s-b)(s-b)$$

Therefore, it is sufficient to prove

$$\frac{abc}{8} \geq (s-a)(s-b)(s-b)$$

Using **Ravi's substitution** (i.e. $a = y + z$, $b = z + x$, $c = x + y$ which can always be done for any arbitrary triangle $\triangle ABC$ with x, y, z strictly positive), the last inequality is equivalent to

$$(x + y)(y + z)(z + x) \geq 8xyz$$

which is straightforward by **AM-GM inequality**

$$(x + y)(y + z)(z + x) \geq 2\sqrt{xy}.2\sqrt{yz}.2\sqrt{zx} = 8xyz$$

as desired. Equality holds for $a = b = c$.

○ **Fifth proof:**

According to **Tereshin-Belabess inequality** *(which we proved in* **Solution 102** *and* **Solution 148**)*, we get*

$$\frac{b^2 + c^2}{4R} \le m_a \le \frac{(b+c)^2}{16r} \le \frac{b^2 + c^2}{8r}$$

Therefore

$$\frac{b^2 + c^2}{4R} \le \frac{b^2 + c^2}{8r}$$

which is equivalent to

$$R \ge 2r$$

as required. The proof is complete.

■

Solution 102.

According to **Tershin's inequality***, we get*

$$\begin{cases} 4Rm_a \ge b^2 + c^2 \\ 4Rm_b \ge c^2 + a^2 \\ 4Rm_c \ge a^2 + b^2 \end{cases}$$

Summing up all these inequalities, we obtain

$$2R(m_a + m_b + m_c) \ge a^2 + b^2 + c^2$$

On the other hand, we know that

$$m_a^2 + m_b^2 + m_c^2 = \frac{3(a^2 + b^2 + c^2)}{4}$$

Therefore

$$\begin{aligned} \frac{m_a^2 + m_b^2 + m_c^2}{m_a + m_b + m_c} &= \frac{3(a^2 + b^2 + c^2)}{4(m_a + m_b + m_c)} \\ &\le \frac{6(m_a + m_b + m_c)R}{4(m_a + m_b + m_c)} \\ &= \frac{3R}{2} \end{aligned}$$

as desired.

The problem is completely solved. Equality holds for $a = b = c$.

Comment 27.

We will give three proofs to **Tereshin's inequality**.

○ **First proof:**

Let s_a and h_a be the lengths of the symmedian and altitude from A. We know that $s_a \geq h_a$ and

$$\begin{cases} s_a = \frac{2bcm_a}{b^2+c^2} \\ h_a = \frac{bc}{2R} \end{cases}$$

Consequently

$$\frac{2bcm_a}{b^2+c^2} \geq \frac{bc}{2R}$$

or equivalently

$$m_a \geq \frac{b^2+c^2}{4R}$$

as required. The proof is complete.

○ **Second proof:**

Let M be the midpoint of $[BC]$ and B' such that $[B'B]$ is a diameter of the circumcircle. If we note by O the circumcenter, then we have

$$2 \cdot OM = B'C = \sqrt{4R^2 - a^2}$$

We need to prove

$$4Rm_a \geq b^2 + c^2$$

which can be transformed into

$$4(m_a - R)^2 \leq 4R^2 - a^2$$

Taking the square root, we get

$$2|m_a - R| \leq B'C$$

or equivalently

$$|AM - AO| \leq OM$$

which is obviously true by the **Triangle inequality**.

○ **Third proof:**

According to the **Sine law**, we obtain

$$\begin{cases} a = 2R\sin A \\ b = 2R\sin B \\ c = 2R\sin C \end{cases}$$

The inequality is equivalent to

$$4(\sin^2 A + \sin^2 B) \leq 4\sqrt{2\sin^2 A + 2\sin^2 B - \sin^2 C}$$

Or

$$(\sin^2 A + \sin^2 B)^2 + \sin^2 C \le 2(\sin^2 A + \sin^2 B)$$

Or

$$\sin^2 C \le (\sin^2 A + \sin^2 B)(\cos^2 A + \cos^2 B)$$

On the other hand, we know that

$$\sin C = \sin A \cos B + \cos A \sin B$$

Therefore, we are left to prove

$$(\sin^2 A + \sin^2 B)(\cos^2 A + \cos^2 B) \ge (\sin A \cos B + \cos A \sin B)^2$$

*which is obviously true by **Cauchy-Schwarz inequality.***

■

Solution 103.

We will prove this inequality in two steps. First of all, we will prove that

$$r_{orthic} = 2R.\cos A.\cos B.\cos C$$

then

$$\cos A \cos B \cos C \le \frac{r^2}{2R^2}$$

○ ***Step 1:***

Let H_A, H_B and H_C be the feet of the altitudes from A, B, C respectively.

Since ABH_AH_B is a cyclic quadrilateral, we have

$$H_AH_B = AB \cos C$$

*Since the circumcircle of $H_AH_BH_C$ is the **nine-point** circle, the circumradius of $H_AH_BH_C$ is just half the circumradius of $\triangle ABC$*

$$R_{orthic} = \frac{R}{2}$$

Let S_{orthic} be the area of the orthic triangle. We have

$$\begin{aligned}
S_{orthic} &= \frac{H_AH_B.H_BH_C.H_CH_A}{4R_{orthic}} \\
&= \frac{H_AH_B.H_BH_C.H_CH_A}{2R} \\
&= \frac{abc}{2R}.\cos A.\cos B.\cos C \\
&= \frac{R^2}{2}.\sin 2A.\sin 2B.\sin 2C
\end{aligned}$$

Similarly, if we note by p_{orthic} the perimeter of the orthic triangle, we get

$$p_{orthic} = R(\sin 2A + \sin 2B + \sin 2C)$$

Consequently

$$r_{orthic} = \frac{2.S_{orthic}}{p_{orthic}}$$

$$= R.\frac{\sin 2A.\sin 2B.\sin 2C}{\sin 2A + \sin 2B + \sin 2C}$$

$$= R.\frac{8\sin A.\sin B.\sin C.\cos A.\cos B.\cos C}{4\sin A.\sin B.\sin C}$$

$$= 2R.\cos A.\cos B.\cos C$$

○ *Step 2:*

Let S be the area of the triangle $\triangle ABC$. We know that

$$\sum_{cyc} a \cdot \cos B.\cos C = \frac{S}{R}$$

Therefore

$$\frac{S}{R\cos A.\cos B.\cos C} = \sum_{cyc} \frac{a}{\cos A}$$

*According to **Cauchy-Schwarz inequality**, we obtain*

$$\frac{S}{R\cos A.\cos B.\cos C} = \sum_{cyc} \frac{a}{\cos A}$$

$$= \sum_{cyc} \frac{a^2}{a\cos A}$$

$$\geq \frac{4s^2}{\sum_{cyc} a\cos A}$$

$$= \frac{4s^2}{4R\sin A.\sin B.\sin C}$$

$$= \frac{2Rs^2}{S}$$

Thus

$$\cos A.\cos B.\cos C \leq \frac{S^2}{2R^2 s^2}$$

$$= \frac{r^2}{2R^2}$$

Finally, we get

$$\frac{r_{orthic}}{r} = \frac{2R.\cos A.\cos B.\cos C}{r}$$

$$\leq \frac{r}{R}$$

$$\leq \frac{1}{2}$$

*where the last step is true by **Euler's inequality**. The proof is complete.*

Solution 104.

We can easily prove that

$$\begin{cases} r_1 = r.\cos A \\ r_2 = r.\cos B \\ r_3 = r.\cos C \end{cases}$$

According to **Solution 103**, *we have*

$$r_4 = 2R.\cos A.\cos B.\cos C$$

○ *Left inequality:*

According to **Solution 101**, *we get*

$$r_1 + r_2 + r_3 + r_4 \geq r_1 + r_2 + r_3$$
$$= r(\cos A + \cos B + \cos C)$$
$$= r\left(1 + \frac{r}{R}\right)$$
$$\geq r$$

○ *Right inequality:*

According to **Solution 101** *and* **Solution 103**, *we have*

$$\begin{cases} \cos A + \cos B + \cos C = 1 + \frac{r}{R} \\ \cos A.\cos B.\cos C \leq \frac{r^2}{2R^2} \end{cases}$$

$$r_1 + r_2 + r_3 + r_4 = r(\cos A + \cos B + \cos C) + 2R.\cos A.\cos B.\cos C$$
$$\leq r\left(1 + \frac{r}{R}\right) + 2R.\frac{r^2}{2R^2}$$
$$= r\left(1 + \frac{2r}{R}\right)$$
$$\leq r(1 + 1)$$
$$= 2r$$

where the last step is true by **Euler's inequality**. *The proof is complete.*

■

Solution 105.

Let A, B, C *be the angles of the triangle* $\triangle ABC$. *We know that*

$$\cos \alpha_H + \cos \beta_H + \cos \gamma_H = 1 + 4\cos A.\cos B.\cos C$$

We need to prove

$$\frac{-4Rr - r^2 + s^2}{R^2} + \frac{5R}{4r} \geq 7$$

*Applying **Blundon's inequality**, we deduce*

$$s^2 \geq 2R^2 + 10Rr - r^2 - 2\sqrt{R(R-2r)^3}$$

Let $x = \sqrt{1 - \frac{2r}{R}} < 1$, *we simply need to prove*

$$\frac{x^2(x^4 + 4x^3 + 3x^2 - 4x + 1)}{2(1-x)(x+1)} \geq 0$$

which is true.

Equality holds for $x = 0$ *which means that the triangle is equilateral.*

∎

Solution 106.

Let A, B, C *be the angles of the triangle and* D, E, F *the feet of the altitudes. If we note by* a, b, c *the side lengths, we have*

$$S_{AEF} = \frac{bc. \cos^2 A. \sin A}{2} = \cos^2 A.S$$

Similarly, we deduce

$$S_{BDE} = \cos^2 B.S$$
$$S_{CDF} = \cos^2 C.S$$

Therefore

$$S_{orthic} = S - S_{AEF} - S_{BDE} - S_{CDF}$$
$$= S\left(1 - \cos^2 A - \cos^2 B - \cos^2 C\right)$$
$$= 2S. \cos A. \cos B. \cos C$$

*According to **Solution 103**, we have*

$$\cos A. \cos B. \cos C \leq \frac{r^2}{2R}$$

Therefore

$$S_{orthic} \leq 2S. \frac{r^2}{2R^2}$$
$$\leq \frac{S}{4}$$

*where the last step is a consequence of **Euler's inequality**.*

The inequality is proved. Equality holds for equilateral triangles.

∎

Solution 107.

We will prove the inequality in two steps. First we will prove

$$a^2 + b^2 + c^2 \geq 8R^2$$

then

$$\frac{a^4 + b^4 + c^4}{a^2 + b^2 + c^2} \geq 3R^2$$

○ *Step 1:*

We need to prove

$$a^2 + b^2 + c^2 \geq 8R^2$$

which is equivalent to

$$2s^2 - 8Rr - 2r^2 \geq 8R^2$$

Or

$$2(s + 2R + r)(s - 2R - r) \geq 0$$

*which is true by **Ciamberlini's inequality.***

○ *Step 2:*

Let AA_1, BB_1, CC_1 be the altitudes of the acute triangle $\triangle ABC$. Thus, $A_1B_1C_1$ is the orthic triangle of ABC for which denote a_0, b_0, c_0 its side-lengths and R_0 , r_0 its circumradius and inradius respectively.

*Apply the well-known inequality weaker than **Gerretsen's inequality***

$$a^2 + b^2 + c^2 \leq 8R^2 + 2Rr$$

to the orthic triangle $A_1B_1C_1$

$$a_0^2 + b_0^2 + c_0^2 \leq 8R_0^2 + 2R_0r_0 \quad (*)$$

Since

$$a_0 = a \cos A \ ; \ b_0 = b \cos B \ ; \ c_0 = c \cos C$$

$$R_0 = \frac{R}{2}$$

$$r_0 = 2R \prod_{cyc} \cos A$$

The inequality $(*)$ becomes

$$\sum_{cyc} a^2 \cdot \cos^2 A \leq 2R^2 + 2R^2 \prod_{cyc} \cos A$$

which is equivalent to

$$\sum_{cyc} a^2 \cdot \left(1 - \frac{a^2}{4R^2}\right) \leq 2R^2 \cdot \left(1 + \frac{a^2 + b^2 + c^2 - 8R^2}{8R^2}\right)$$

Or

$$\sum_{cyc} a^2 \cdot (4R^2 - a^2) \le R^2 \cdot (a^2 + b^2 + c^2)$$

Or

$$\sum_{cyc} a^4 \ge 3R^2 \left(\sum_{cyc} a^2 \right)$$

Therefore

$$\frac{a^4 + b^4 + c^4}{a^2 + b^2 + c^2} \ge 3R^2$$

Equality holds if and only if the triangle is equilateral.

Comment 28.

We can prove **Gerresten's inequality** the same way we proved **Blundon's inequality**. In fact, we have

$$(a - b)^2 (b - c)^2 (c - a)^2 = -4r^2 [(s^2 - 2R^2 - 10Rr + r^2)^2 - 4R(R - 2r)^3]$$

○ **Right inequality:**

From the previous identity

$$s^2 \le 2R^2 + 10Rr - r^2 + 2(R - 2r)\sqrt{R(R - 2r)}$$

$$= 2R^2 + 10Rr - r^2 + (R - 2r)\frac{2(R - r)\sqrt{R(R - 2r)}}{R - r}$$

$$\le 2R^2 + 10Rr - r^2 + (R - 2r)\frac{(R - r)^2 + R(R - 2r)}{R - r}$$

$$= 4R^2 + 4Rr + 3r^2 - \frac{r^2(R - 2r)}{R - r}$$

$$\le 4R^2 + 4Rr + 3r^2$$

○ **Left inequality:**

In a similar way, we get

$$s^2 \ge 2R^2 + 10Rr - r^2 - 2(R - 2r)\sqrt{R(R - 2r)}$$

$$= 2R^2 + 10Rr - r^2 - (R - 2r)\frac{2(R - r)\sqrt{R(R - 2r)}}{R - r}$$

$$\ge 2R^2 + 10Rr - r^2 - (R - 2r)\frac{(R - r)^2 + R(R - 2r)}{R - r}$$

$$= 16Rr - 5r^2 + \frac{r^2(R - 2r)}{R - r}$$

$$\ge 16Rr - 5r^2$$

Combining the two inequalities, we get

$$16Rr - 5r^2 \le s^2 \le 4R^2 + 4Rr + 3r^2$$

The proof is now complete.

Solution 108.

We will give two solutions to this inequality.

• **First Solution:**

Let $z = a^2 + b^2 - c^2$, $y = c^2 + a^2 - b^2$ *and* $x = b^2 + c^2 - a^2$. *Thus,* x, y, z *are positive real numbers, and we need to prove that*

$$\sum_{cyc} \sqrt{(y+z)(4x+y+z)} \leq \frac{9\sqrt{3}(x+y)(y+z)(z+x)}{4(xy+yz+zx)}$$

*Using **Cauchy-Schwarz inequality**, we find that*

$$\sum_{cyc} \sqrt{(y+z)(4x+y+z)} \leq \sqrt{\left(\sum_{cyc}(x+y)\right)\left(\sum_{cyc}(4x+y+z)\right)} = 2\sqrt{3}(x+y+z)$$

It is sufficient to prove

$$9(x+y)(y+z)(z+x) \geq 8(x+y+z)(xy+yz+zx)$$

which is equivalent to

$$\sum_{cyc} x(y-z)^2 \geq 0$$

The proof is complete. Equality holds for $a = b = c$.

• **Second solution:**

*According to **Cauchy-Schwarz inequality**, we get*

$$am_a + bm_b + cm_c \leq \sqrt{(a^2 + b^2 + c^2)(m_a^2 + m_b^2 + m_c^2)}$$

On the other hand, we know that

$$m_a^2 + m_b^2 + m_c^2 = \frac{3}{4}(a^2 + b^2 + c^2)$$

Therefore

$$am_a + bm_b + cm_c \leq \frac{\sqrt{3}}{2}(a^2 + b^2 + c^2)$$

*But, we know that (the details are in **Comment 29**)*

$$a^2 + b^2 + c^2 \leq 9R^2$$

Consequently

$$am_a + bm_b + cm_c \leq \frac{9\sqrt{3}R^2}{2}$$

This ends the proof. Equality holds for $a = b = c$.

Comment 29.

We will show here how to prove

$$a^2 + b^2 + c^2 \leq 9R^2$$

In fact, by squaring $\overrightarrow{BC} = \overrightarrow{OC} - \overrightarrow{OB}$, *we get*

$$a^2 = 2R^2 - 2\overrightarrow{OB}.\overrightarrow{OC}$$

Or

$$2.\overrightarrow{OB}.\overrightarrow{OC} = 2R^2 - a^2$$

Similarly

$$2.\overrightarrow{OA}.\overrightarrow{OC} = 2R^2 - b^2$$
$$2.\overrightarrow{OA}.\overrightarrow{OB} = 2R^2 - c^2$$

It follows that

$$9.OG^2 = \|\overrightarrow{OA} + \overrightarrow{OB} + \overrightarrow{OC}\|^2$$
$$= 3R^2 + 2\left(\overrightarrow{OA}.\overrightarrow{OB} + \overrightarrow{OB}.\overrightarrow{OC} + \overrightarrow{OC}.\overrightarrow{OA}\right)$$
$$= 9R^2 - (a^2 + b^2 + c^2) \geq 0$$

as required.

■

Solution 109.

Let

$$\begin{cases} x = \frac{b^2 + c^2 - a^2}{2} \geq 0 \\ y = \frac{c^2 + a^2 - b^2}{2} \geq 0 \\ z = \frac{a^2 + b^2 - c^2}{2} \geq 0 \end{cases}$$

Therefore

$$\begin{cases} a = \sqrt{y + z} \\ b = \sqrt{z + x} \\ c = \sqrt{x + y} \end{cases}$$

WLOG, we can assume $x \geq y \geq z$. *According to* **Heron's formula**, *we get*

$$S_{ABC} = \frac{1}{2}\sqrt{xy + yz + zx}$$

Thus

$$\max(h_a, h_b, h_c) = \frac{2S_{ABC}}{a} = \frac{\sqrt{xy + yz + zx}}{\sqrt{y + z}}$$

On the other hand, we have

$$R = \frac{abc}{4S_{ABC}} = \frac{\sqrt{(x + y)(y + z)(z + x)}}{2\sqrt{xy + yz + zx}}$$

Thus, we need to prove

$$2(xy + yz + zx)^2 \geq (y+z)^2(x+y)(x+z)$$

Consider the following quadratic

$$f(x) = 2(xy + yz + zx)^2 - (y+z)^2(x+y)(x+z)$$

We have

$$f(x) - f(y) = (x-y)(y+z)(xy + xz + 3yz - z^2) \geq 0$$

Therefore

$$f(x) \geq f(y)$$
$$= 2y^2(y + 2z)^2 - 2y(y+z)^3$$
$$= 2yz(y^2 + yz - z^2)$$
$$\geq 0$$

The proof is complete.

■

Solution 110.

First of all, we will prove

$$p_2 = \frac{2S_{ABC}}{R}$$

where S_{ABC} is the area, and R is radius of circumscribed circle of triangle $\triangle ABC$.

In fact, we have

$$p_2 = \sum_{cyc} a \cos A$$
$$= \sum_{cyc} 2R. \sin A. \cos A$$
$$= R \left(\sum_{cyc} \sin 2A \right)$$
$$= 4R \prod_{cyc} \sin A$$
$$= \frac{abc}{2R^2}$$
$$= \frac{2S_{ABC}}{R}$$

Our lemma is proved.

We will prove each side separately as follows:

○ **Right inequality:**

Let s and r be semiperimeter and inradius of the triangle $\triangle ABC$, respectively. Applying **Euler's formula**, we get

$$OI = \sqrt{R(R - 2r)}$$

We need to prove

$$2\sqrt{R(R - 2r)} \geq s - \frac{2sr}{R}$$

$$\Longleftrightarrow 2R\sqrt{R(R - 2r)} \geq s(R - 2r)$$

$$\Longleftrightarrow 4R^3 \geq s^2(R - 2r)$$

According to **Gerretsen's inequality**, it is enough to prove

$$4R^3 \geq (4R^2 + 4Rr + 3r^2)(R - 2r)$$

or equivalently

$$r(6r^2 + 5rR + 4R^2) \geq 0$$

which is obviously true.

○ **Left inequality:**

We need to prove

$$\frac{p_1}{2} - p_2 \geq 0$$

which is equivalent to prove

$$\frac{S_{ABC}}{r} - \frac{2S_{ABC}}{R} \geq 0$$

Or

$$R \geq 2r$$

which is just **Euler's inequality**.

Comment 30.

We will show here how to prove **Euler's formula**.

Let D be the tangency point of the incircle with AB, and F the intersection of AI with the circumcirle.

Applying the **Power Of Point** I, we get

$$R^2 - OI^2 = IA.IF$$

On the other hand, we have

$$AI = \frac{r}{\sin \frac{\angle A}{2}}$$

Furthermore

$$\angle BIF = \angle IBF = \frac{\angle A + \angle B}{2}$$

Thus, the triangle $\triangle IBF$ is isosceles, and we have

$$IF = BF = 2R\sin\frac{\angle A}{2}$$

Therefore

$$IA.IF = 2Rr = R^2 - OI^2$$

and the result follows.

∎

Solution 111.

*We will use the following formula which we proved in **Solution 110***

$$p_{orthic} = \frac{2sr}{R}$$

○ **Right inequality:**

We need to prove

$$\frac{2sr}{R} \leq 3\sqrt{3}r$$

which is equivalent to

$$27R^2 \geq 4s^2$$

*According to **Gerretsen's inequality**, it is sufficient to prove*

$$27R^2 \geq 4(4R^2 + 4Rr + 3r^2)$$

which is equivalent to

$$(6r + 11R)(R - 2r) \geq 0$$

*This last inequality is obviously true by **Euler's inequality**.*

○ **Left inequality:**

The inequality is equivalent to prove

$$a + b + c > 4R$$

Since the circumcenter is the intersection of the perpendicular bisectors of a triangle, for any acute triangle, it will lie within the medial triangle (triangle whose vertices are the midpoints of the sides of ABC). Let M be the midpoint of AB, N the midpoint of AC, and O the circumcenter of ABC. Therefore, we have

$$BM + MN + NC > BO + OC$$

$$\Longleftrightarrow \frac{a}{2} + \frac{b}{2} + \frac{c}{2} > 2R$$

$$\Longleftrightarrow a + b + c > 4R$$

The inequality is proved.

Solution 112.

We have

$$
\begin{cases}
BP = \frac{ac}{b+c} \\
BQ = \frac{ac}{a+b}
\end{cases}
$$

Therefore

$$
\begin{aligned}
S_{PBQ} &= \frac{a^2 c^2}{2(a+b)(b+c)} \sin B \\
&= \frac{ac.S_{ABC}}{(a+b)(b+c)}
\end{aligned}
$$

Similarly, we get

$$
S_{QAR} = \frac{bc.S_{ABC}}{(a+b)(a+c)}
$$

$$
S_{RCP} = \frac{ab.S_{ABC}}{(a+c)(b+c)}
$$

Thus

$$
\begin{aligned}
S_{PQR} &= S_{ABC} - S_{PBQ} - S_{QAR} - S_{RCP} \\
&= S_{ABC} \left(1 - \frac{ac}{(a+b)(b+c)} - \frac{bc}{(a+b)(a+c)} - \frac{ab}{(a+c)(b+c)} \right) \\
&= \frac{2abc}{(a+b)(b+c)(a+c)} . S_{ABC}
\end{aligned}
$$

It is enough to prove

$$
(a+b)(b+c)(c+a) \geq 8abc
$$

which is well-known and easy to prove by **AM-GM inequality**

$$
(a+b)(b+c)(c+a) \geq 2\sqrt{ab}.2\sqrt{bc}.2\sqrt{ca}
$$
$$
= 8abc
$$

The inequality is proved.

∎

Solution 113.

We will give two solutions to this problem.

● **First Solution:**

By **Cauchy-Schwarz inequality**, we find that

$$
\sum_{cyc} \frac{\cos^3 A}{\sin B \sin C} \geq \frac{\left(\cos^2 A + \cos^2 B + \cos^2 C \right)^2}{\sum_{cyc} \cos A \sin B \sin C} = \frac{\left(\cos^2 A + \cos^2 B + \cos^2 C \right)^2}{\frac{1}{2} \left(\sin^2 A + \sin^2 B + \sin^2 C \right)}
$$

Thus, it remains to prove that

$$\left(\cos^2 A + \cos^2 B + \cos^2 C\right)^2 \geq \sin^2 A + \sin^2 B + \sin^2 C$$

or equivalently

$$(4(\cos^2 A + \cos^2 B + \cos^2 C) - 3))(\cos^2 A + \cos^2 B + \cos^2 C + 1) \geq 0$$

*which can easily be proved using the result from **Solution 103** as follows*

$$\cos^2 A + \cos^2 B + \cos^2 C = 1 - 2\cos A \cos B \cos C \geq 1 - \frac{r^2}{R^2} \geq \frac{3}{4}$$

The proof is complete.

- *Second Solution:*

With $a = 2\cos A$, $b = 2\cos B$, $c = 2\cos C$, *the original problem is equivalent to the following* $a^2 + b^2 + c^2 + abc = 4$ *and we need to prove*

$$\sum_{cyc} \frac{a^3}{\sqrt{4 - b^2}\sqrt{4 - c^2}} \geq 1$$

Or

$$2a + bc = \sqrt{4 - b^2}\sqrt{4 - c^2}$$

Therefore, we need to prove

$$\sum_{cyc} \frac{a^3}{2a + bc} \geq 1$$

*By **Cauchy-Schwarz inequality**, we find that*

$$\sum_{cyc} \frac{a^3}{2a + bc} = \sum_{cyc} \frac{a^4}{2a^2 + abc} \geq \frac{(a^2 + b^2 + c^2)^2}{2(a^2 + b^2 + c^2) + 3abc} = \frac{(a^2 + b^2 + c^2)^2}{12 - a^2 - b^2 - c^2}$$

It is enough to prove

$$\frac{(a^2 + b^2 + c^2)^2}{12 - a^2 - b^2 - c^2} \geq 1$$

or

$$(a^2 + b^2 + c^2 - 3)(a^2 + b^2 + c^2 + 4) \geq 0$$

which is obviously true as $a^2 + b^2 + c^2 \geq 3$. *Equality holds for* $a = b = c = 1$.

■

Solution 114.

The original inequality is equivalent to

$$189R^2 \geq 25(a^2 + b^2 + c^2) - 4(ab + bc + ca)$$

Using sRr notations, we can rewrite the inequality in the following form

$$189R^2 \geq 46s^2 - 216Rr - 54r^2$$

or equivalently

$$(5R + 42r)(R - 2r) + 46(4R^2 + 4Rr + 3r^2 - s^2) \geq 0$$

*which is obviously true according to **Euler's** and **Gerretsen's** inequalities.*

The proof is complete.

Solution 115.

We will prove that the minimum is 4. The inequality is equivalent to

$$\frac{1}{a^2} + \frac{1}{b^2} + \frac{1}{c^2} + \frac{\sum\limits_{cyc}(2a^2b^2 - a^4)}{3a^2b^2c^2} \geq \frac{36}{(a+b+c)^2}$$

*which is $f(w^3) \geq 0$, where f is a concave function. According to **uvw method**, it is enough to prove our inequality in the following cases:*

(i) **The first case:** $b = 1$ *and* $c = 1 + a$.

In this case, we get

$$(a^2 + a - 2)^2(a^2 + 4a + 1) \geq 0$$

(ii) **The second case:** $b = c = 1$ *and* $0 < a < 2$.

In this case, we get

$$(a - 1)^2(12 + 36a - 5a^2 - 6a^3 - a^4) \geq 0$$

which is true.

We proved the inequality in all cases.

■

Solution 116.

Using sRr notations, we get

$$s = \frac{a + b + c}{2} = \frac{3}{2}$$

We will use the following inequality

$$16Rr - 5r^2 + \frac{r^2(R - 2r)}{R - r} \leq s^2$$

*Let $x = \frac{R}{r}$ and by **Euler's** inequality, we have $x \geq 2$. Moreover*

$$\frac{1}{R^2} + \frac{1}{r^2} - \frac{6R}{r} \geq \frac{4}{9}\left(16Rr - 5r^2 + \frac{r^2(R - 2r)}{R - r}\right)\left(\frac{1}{R^2} + \frac{1}{r^2}\right) - \frac{6R}{r}$$

$$= 3 + \frac{(x - 2)(10x^3 - 33x^2 + 37x - 6)}{9(x - 1)x^2}$$

$$\geq 3$$

Equality at $x = 2$, meaning the triangle is equilateral.

Solution 117.

We will prove each side separately.

- ### Right inequality:

*Let A' be the point in the plan such that $ABA'C$ is a parallelogram. According to the **Triangle inequality**, we have*

$$2m_a = AA' < b + c$$

Similarly

$$2m_b < c + a$$

$$2m_c < a + b$$

Adding up all these inequalities, we get

$$\frac{m_a + m_b + m_c}{a + b + c} < 1$$

- ### Left inequality:

We will prove this inequality in two different ways.

- ### First Solution:

Let m'_a, m'_b, m'_c be the medians of the triangle formed by m_a, m_b and m_c. Applying the previous result, we get

$$m'_a + m'_b + m'_c < m_a + m_b + m_c$$

*On the other hand, according to **Apollonius' theorem**, we get*

$$m'_a = \frac{1}{2}\sqrt{2(m_b^2 + m_c^2) - m_a^2} = \frac{3a}{4}$$

Similarly

$$m'_b = \frac{3b}{4}$$

$$m'_c = \frac{3c}{4}$$

Therefore

$$\frac{3}{4}(a + b + c) < m_a + m_b + m_c$$

or equivalently

$$\frac{3}{4} < \frac{m_a + m_b + m_c}{a + b + c}$$

The proof is complete.

○ *Second Solution:*

If we note by G the centroid of ABC, we have

$$\frac{2}{3}(m_a + m_b) = GA + GB > AB = c$$

Similarly

$$\frac{2}{3}(m_b + m_c) = GB + GC > BC$$

and

$$\frac{2}{3}(m_a + m_c) = GA + GC > AC$$

Summing up all the inequalities, we get

$$\frac{3}{4} < \frac{m_a + m_b + m_c}{a + b + c}$$

Finally, combining the two results, we get that

$$\frac{3}{4} < \frac{m_a + m_b + m_c}{a + b + c} < 1$$

The proof is complete.

∎

Solution 118.

○ *Left inequality:*

Let h_a, h_b, h_c be the altitudes, and S the area of the triangle. We have

$$m_a + m_b + m_c \geq h_a + h_b + h_c$$

It is sufficient to prove

$$9r \leq h_a + h_b + h_c$$

which is true because

$$\left(\frac{1}{h_a} + \frac{1}{h_b} + \frac{1}{h_c}\right)(h_a + h_b + h_c) \geq 3\sqrt[3]{h_a h_b h_c} \, 3\sqrt[3]{\frac{1}{h_a h_b h_c}}$$

$$= 9$$

and

$$\frac{1}{h_a} + \frac{1}{h_b} + \frac{1}{h_c} = \frac{a}{2S} + \frac{b}{2S} + \frac{c}{2S}$$

$$= \frac{a + b + c}{2S}$$

$$= \frac{1}{r}$$

○ *Right inequality:*

This inequality is known as **Leuenberger's inequality** and we will prove it in two different ways.

- *First Solution:*

Let d_a, d_b, d_c be the distances from the circumcenter to the corresponding side lines. According to the **Triangle inequality**, we obtain

$$\begin{cases} m_a \le R + d_a \\ m_b \le R + d_b \\ m_c \le R + d_c \end{cases}$$

Therefore

$$m_a + m_b + m_c \le 3R + d_a + d_b + d_c$$

On the other hand

$$\begin{cases} d_a = R\cos A \\ d_b = R\cos B \\ d_c = R\cos C \end{cases}$$

According to **Solution 101**, we obtain

$$\cos A + \cos B + \cos C = 1 + \frac{r}{R}$$

Consequently

$$\begin{aligned} m_a + m_b + m_c &\le 3R + d_a + d_b + d_c \\ &= 3R + R(\cos A + \cos B + \cos C) \\ &= 4R + r \end{aligned}$$

as required.

- *Second Solution:*

Let M and N be the midpoints of $[BC]$ and $[AC]$ respectively. According to **Ptolemy's inequality**, we get

$$\begin{aligned} m_a.m_b &= AM.BN \\ &\le AB.MN + AN.BM \\ &= c.\frac{c}{2} + \frac{a}{2}.\frac{b}{2} \\ &= \frac{2c^2 + ab}{4} \end{aligned}$$

Therefore

$$4m_a m_b \le 2c^2 + ab$$

(Alternatively: $(2c^2 + ab)^2 - (4m_a m_b)^2 = (a-b)^2(a+b-c)(a+b+c) \geq 0)$

After squaring both sides and using $4m_a m_b \leq 2c^2 + ab$, *we need to prove that*

$$\sum_{cyc}(a^6 - a^4b^2 - a^4c^2 - a^4bc + a^3b^2c + a^3c^2b) \geq 0$$

that we can rewrite in the following form

$$(a+b+c)\left(\sum_{cyc} a^3(a-b)(a-c)\right) \geq 0$$

which is true by **fifth degree Schur's inequality***.*

The proof is complete.

■

Solution 119.

We will give two solutions to this problem.

● *First Solution:*

According to **Apollonius' theorem***, we get*

$$\begin{cases} m_a = \frac{1}{2}\sqrt{2b^2 + 2c^2 - a^2} \\ m_b = \frac{1}{2}\sqrt{2c^2 + 2a^2 - b^2} \\ m_c = \frac{1}{2}\sqrt{2a^2 + 2b^2 - c^2} \end{cases}$$

Therefore

$$\sum_{cyc}\frac{m_a}{a} = \frac{1}{2}\sum_{cyc}\sqrt{\frac{2b^2 + 2c^2 - a^2}{a^2}}$$

$$= \frac{1}{\sqrt{3}}\sum_{cyc}\frac{1}{2\sqrt{\frac{a^2}{2b^2+2c^2-a^2}\cdot\frac{1}{3}}}$$

$$\geq \frac{1}{\sqrt{3}}\sum_{cyc}\frac{1}{\frac{a^2}{2b^2+2c^2-a^2}+\frac{1}{3}}$$

$$= \sum_{cyc}\frac{\sqrt{3}(2b^2 + 2c^2 - a^2)}{2(a^2 + b^2 + c^2)}$$

$$= \frac{3\sqrt{3}}{2}$$

The proof is complete. Equality holds for $a = b = c$*.*

• *Second Solution:*

Let $x = a^2$, $y = b^2$ and $z = c^2$. The lengths of the medians can be obtained from *Apollonius' theorem* as

$$
\begin{cases}
m_a = \sqrt{\frac{2b^2+2c^2-a^2}{4}} = \sqrt{\frac{2y+2z-x}{4}} \\
m_b = \sqrt{\frac{2c^2+2a^2-b^2}{4}} = \sqrt{\frac{2z+2x-y}{4}} \\
m_c = \sqrt{\frac{2a^2+2b^2-c^2}{4}} = \sqrt{\frac{2x+2y-z}{4}}
\end{cases}
$$

Therefore, we need to prove

$$
\sum_{cyc} \sqrt{\frac{2y+2z-x}{x}} \geq 3\sqrt{3}
$$

Applying **Hölder's inequality**, we obtain

$$
\left(\sum_{cyc} \sqrt{\frac{2y+2z-x}{x}} \right)^2 \left(\sum_{cyc} x(2y+2z-x)^2 \right) \geq 27(x+y+z)^3
$$

Thus, it is enough to prove that

$$
(x+y+z)^3 \geq \sum_{cyc} x(2y+2z-x)^2
$$

which is equivalent to

$$
\sum_{sym} x^2 y \geq 6xyz
$$

This last inequality is obviously true by **AM-GM inequality**.

∎

Solution 120.

First of all, we have

$$
\sum_{cyc} \frac{h_a}{b+c} = \frac{1}{2R} \left(\sum_{cyc} \frac{bc}{b+c} \right)
$$

Using **Power Mean inequality**, we get

$$
\frac{bc}{b+c} \leq \frac{b+c}{4}
$$

Therefore

$$
\sum_{cyc} \frac{h_a}{b+c} = \frac{1}{2R} \left(\sum_{cyc} \frac{bc}{b+c} \right)
$$
$$
\leq \sum_{cyc} \frac{b+c}{8R}
$$

$$= \frac{a+b+c}{4R}$$

$$= \frac{\sin A + \sin B + \sin C}{2}$$

$$\leq \frac{3}{2} \sin \left(\frac{A+B+C}{3} \right)$$

$$= \frac{3}{2} \sin \left(\frac{\pi}{3} \right)$$

$$= \frac{3\sqrt{3}}{4}$$

where the last inequality is true by **Jensen's inequality**.

∎

Solution 121.

○ *Left inequality:*

First of all, we know that

$$\frac{1}{h_a} + \frac{1}{h_b} + \frac{1}{h_c} = \frac{1}{r}$$

Applying **AM-GM inequality**, we get that

$$(h_a + h_b + h_c) \left(\frac{1}{h_a} + \frac{1}{h_b} + \frac{1}{h_c} \right) \geq 9$$

Therefore

$$h_a + h_b + h_c \geq 9r$$

○ *Right inequality:*

We can rewrite the inequality using sRr notations, and we get

$$2r(R - 2r) + (4R^2 + 4rR + 3r^2 - s^2) \geq 0$$

which is true by **Euler's** and **Gerretsen's inequalities**.

∎

Solution 122.

Let S be the area of the triangle. We have

$$h_a + h_b + h_c = 2S \left(\frac{1}{a} + \frac{1}{b} + \frac{1}{c} \right)$$

According to **Heron's Formula**, we obtain

$$S = \frac{1}{2} \sqrt{(a+b+c)(a+b-c)(b+c-a)(c+a-b)}$$

Therefore, it is enough to prove

$$3(a+b+c)a^2b^2c^2 \geq (ab+bc+ca)^2(a+b-c)(a+c-b)(b+c-a)$$

On the other hand, we know that

$$a^2b^2 + b^2c^2 + c^2a^2 \geq \frac{(ab+bc+ca)^2}{3}$$

We will prove the following stronger result

$$(a+b+c)a^2b^2c^2 \geq (a^2b^2 + b^2c^2 + c^2a^2)(a+b-c)(a+c-b)(b+c-a)$$

which is equivalent to

$$\sum_{cyc} a^2b^2(a-b)^2(a+b-c) \geq 0$$

This last inequality is obviously true.

∎

Solution 123.

WLOG, we can assume that $c \leq b \leq a$. Then, we have

$$\min(m_a, m_b, m_c) = m_a = \frac{\sqrt{2b^2 + 2c^2 - a^2}}{2}$$

$$\max(h_a, h_b, h_c) = h_c = \frac{2S}{c}$$

Therefore, we need to prove

$$\sqrt{2b^2 + 2c^2 - a^2} \geq \frac{2S}{c}$$

Or

$$c^2(2b^2 + 2c^2 - a^2) \geq \frac{(a+b+c)(-a+b+c)(a-b+c)(a+b-c)}{4}$$

Or

$$4c^2(2b^2 + 2c^2 - a^2) \geq -a^4 - b^4 - c^4 + 2a^2b^2 + 2a^2c^2 + 2b^2c^2$$

Or

$$a^4 + b^4 + 9c^4 + 6b^2c^2 \geq 2a^2b^2 + 6a^2c^2$$

$$(a^2 - b^2 - 3c^2)^2 \geq 0$$

which is obviously true.

∎

Solution 124.

Let p_{median} and A_{median} be the perimeter and area of the triangle formed by the lengths of the medians respectively. It is well known that

$$A_{medians} = \frac{3A}{4}$$

*and according to **Problem 117**, we have*

$$\frac{3}{4}p < p_{median} < p$$

On the other hand, the inradius r is

$$r = \frac{2A}{a+b+c}$$

Thus

$$\frac{r_{medians}}{r} = \frac{\frac{3A}{2}}{\frac{p_{medians}}{\frac{2A}{p}}} = \frac{3p}{4p_{medians}}$$

Therefore

$$\frac{r_{medians}}{r} \geq \frac{3p}{4p} = \frac{3}{4}$$

Similarily, we have

$$\frac{r_{medians}}{r} \leq \frac{3p}{4 \times \frac{3p}{4}} = 1$$

The proof is complete.

■

Solution 125.

After some simplifications, we need to prove

$$\sum_{cyc}(a^4 - a^3b - a^3c + a^2bc) \geq 0$$

*which is just **fourth degree Schur's inequality***

$$\sum_{cyc}(a^4 - a^3b - a^3c + a^2bc) = \sum_{cyc}a^2(a-b)(a-c) \geq 0$$

as required.

■

Solution 126.

WLOG, let $b > c$, as otherwise the case $b = c$ is trivial.

Now simple length chasing gives the fact that the foot of A on BC, the incircle touchpoint with BC, the foot of the $A-$angle bisector and the midpoint of BC lie in this order. So we have

$$m_a - l_a \geq 0$$

For the second part, note that the difference is less than the distance between the midpoint of BC and the foot of the $A-$angle bisector. This distance is

$$\frac{a}{2} - \frac{ac}{b+c} = \frac{a(b-c)}{2(b+c)} < \frac{b-c}{2}$$

so we are done.

Comment 31.

We aslo have the following identity

$$4(b+c)^2(m_a^2 - l_a^2) = (b-c)^2(2(b+c)^2 - a^2) \geq 0$$

∎

Solution 127.

Let the incircle be tangent to BC at X, and the midpoint of BC be M. Therefore

$$2OI \geq 2MX = |b - c|$$

and the inequality is proved.

∎

Solution 128.

Notice that the difference between the median and the interior bisector is less than the distance between the midpoint and the foot of the $A-$angle bisector. This distance is

$$\left| \frac{a}{2} - \frac{ac}{b+c} \right| = \frac{a|b-c|}{2(b+c)} < \frac{|b-c|}{2}$$

Therefore

$$m_a - l_a \leq \frac{|b-c|}{2}$$

and similarly for $m_b - l_b$ and $m_c - l_c$.

Finally

$$\sum_{cyc}(m_c - l_c)^2 \leq \sum_{cyc}\frac{c^2(a-b)^2}{4(a+b)^2} \leq \frac{1}{4}\sum_{cyc}(a-b)^2$$

The proof is complete.

Solution 129.

After using **Ravi's substitution**, we need to prove

$$\sum_{cyc}(x^2 - xy) \leq \frac{\prod_{cyc}(x+y)\sum_{cyc}z(x-y)^2}{4xyz(x+y+z)}$$

Now assuming that $xyz(x+y+z) = 3$, we need to prove

$$12\sum_{cyc}(x^2 - xy) \leq \prod_{cyc}(x+y)\sum_{cyc}z(x-y)^2$$

which is exactly **Problem 48**.

∎

Solution 130.

We will use sRr notations. Applying **Euler's formula**, we get

$$OI^4 = R^2(R - 2r)^2$$

and

$$\sum_{cyc}(a-b)^4 = 2(12Rr + 3r^2 - s^2)^2$$

Therefore we will show that

$$16R^2(R - 2r)^2 \geq (12Rr + 3r^2 - s^2)^2$$

which is equivalent to

$$2(4R^2 + 4Rr + 3r^2 - s^2)\left[4(R+r)(R-2r) + (s^2 + 5r^2 - 16Rr)\right] \geq 0$$

which is true because of the following famous results

$$R \geq 2r$$

$$4R^2 + 4Rr + 3r^2 \geq s^2$$

$$s^2 + 5r^2 \geq 16Rr + \frac{r^2(R-2r)}{R-r} \geq 16Rr$$

The proof is complete.

∎

Solution 131.

We need to prove that

$$\sum_{cyc}\frac{m_a}{a}(b+c-2a) \geq 0$$

which is true by **Chebyshev's inequality**

$$\sum_{cyc}\frac{m_a}{a}(b+c-2a) \geq \frac{1}{3}\left(\sum_{cyc}\frac{m_a}{a}\right)\left(\sum_{cyc}(b+c-2a)\right) = 0$$

Solution 132.

First of all, we can easily prove the following equality

$$\frac{a}{b+c} + \frac{b}{c+a} + \frac{c}{a+b} = \frac{2(s^2 - Rr - r^2)}{s^2 + 2Rr + r^2}$$

Consequently, it is sufficient to prove

$$\frac{2(s^2 - Rr - r^2)}{s^2 + 2Rr + r^2} \le \frac{R}{6r} + \frac{7}{6}$$

Let

$$f(s^2) = \frac{R}{6r} + \frac{7}{6} - \frac{2(s^2 - Rr - r^2)}{s^2 + 2Rr + r^2}$$

f is a decreasing function and we need to prove

$$f(s^2) \ge 0$$

*Applying **Gerretsen's inequality**, we deduce that*

$$s^2 \le 4R^2 + 4Rr + 3r^2$$

Therefore, it is sufficient to prove

$$f(4R^2 + 4Rr + 3r^2) \ge 0$$

After some simplifications, we get the following inequality

$$(R - 2r)\left(2(R - 2r)r + 5(R - 2r) + r^2\right) \ge 0$$

*which is obviously true by **Euler's inequality**.*

■

Solution 133.

We have

$$\sum_{cyc} m_a^2 = \frac{3}{4}(a^2 + b^2 + c^2) = \frac{3p^2 - 12Rr - 3r^2}{2},$$

Therefore

$$\frac{27R^2}{4} - \sum_{cyc} m_a^2 = \frac{3}{4}(R + 2r)(R - 2r) + \frac{3}{2}(4R^2 + 4Rr + 3r^2 - s^2) \ge 0$$

*which is true by **Euler's** and **Gerretsen's** inequalities*

Similarly, we can prove the other side of the inequality

$$\sum_{cyc} m_a^2 - 6R^2 = \frac{3}{2}(2R + p + r)(s - 2R - r) \ge 0$$

*which is true by **Ciamberlini's inequality**.*

Solution 134.

We will give two solutions to this problem.

• *First Solution:*

Since the circumcenter is the intersection of the perpendicular bisectors of a triangle, for any acute triangle, it will lie within the medial triangle (triangle whose vertices are the midpoints of the sides of ABC). Let M be the midpoint of AB, N the midpoint of AC, and O the circumcenter of ABC. Then, we obviously have

$$BM + MN + NC > BO + OC$$

Or

$$\frac{a}{2} + \frac{b}{2} + \frac{c}{2} > 2R$$

Or

$$a + b + c > 4R$$

*According to **Euler's formula**, we have*

$$OI = \sqrt{R^2 - 2Rr} \le R$$

Therefore

$$a + b + c > 4OI$$

as desired.

• *Second Solution:*

Note that $a + b + c = 2s$ and $OI^2 = R^2 - 2Rr$. Therefore

$$(a + b + c)^2 - (4OI)^2 = 4s^2 + 32Rr - 16R^2$$
$$= 48r(R - 2r) + (8R + 4s + 4r)(s - 2R - r) + 100r^2$$
$$> 0$$

*which is true by **Euler's** and **Ciamberlini's** inequalities.*

■

Solution 135.

*Applying **Euler's formula**, we obtain*

$$OI^2 = R^2 - 2Rr$$

and

$$\sum_{cyc}(a - b)^2 = 2s^2 - 24Rr - 6r^2$$

*According to **Walker's inequality**, we get*

$$\sum_{cyc}(a - b)^2 - 4.OI^2 = 2(s^2 - 2R^2 - 8Rr - 3r^2) \ge 0$$

According to **Gerretsen's inequality**, we obtain

$$8.OI^2 - \sum_{cyc}(a-b)^2 = 2(4R^2 + 4Rr + 3r^2 - s^2) \geq 0$$

as required.

∎

Solution 136.

○ *Right inequality:*

Let m_a, m_b, m_c be the medians of the triangle. We know that $m_a \geq h_a$, $m_b \geq h_b$ and $m_c \geq h_c$, therefore

$$\frac{h_a^2 + h_b^2 + h_c^2}{a^2 + b^2 + c^2} \leq \frac{m_a^2 + m_b^2 + m_c^2}{a^2 + b^2 + c^2}$$

On the other hand, we know that

$$m_a^2 + m_b^2 + m_c^2 = \frac{3(a^2 + b^2 + c^2)}{4}$$

Consequently

$$\frac{h_a^2 + h_b^2 + h_c^2}{a^2 + b^2 + c^2} \leq \frac{3}{4}$$

○ *Left inequality:*

Let D be the foot of the altitude from A. Then

$$h_a^2 = \frac{b^2 + c^2 - DB^2 - DC^2}{2}$$
$$> \frac{b^2 + c^2 - (DB + DC)^2}{2}$$
$$= \frac{b^2 + c^2 - a^2}{2}$$

and similarly for B and C. Therefore

$$h_a^2 + h_b^2 + h_c^2 > \sum_{cyc} \frac{b^2 + c^2 - a^2}{2}$$
$$= \frac{a^2 + b^2 + c^2}{2}$$

This ends the proof.

∎

Solution 137.

○ **Right inequality:**

If we note by m_a, m_b, m_c the medians of the triangle, we have the following inequality

$$4(b+c)^2(m_a^2 - l_a^2) = (b-c)^2(2(b+c)^2 - a^2) \geq 0$$

Similarly, we can prove $m_b \geq l_b$ and $m_c \geq l_c$. Therefore

$$\frac{l_a^2 + l_b^2 + l_c^2}{a^2 + b^2 + c^2} \leq \frac{m_a^2 + m_b^2 + m_c^2}{a^2 + b^2 + c^2} = \frac{3}{4}$$

○ **Left inequality:**

We need to prove

$$2(l_a^2 + l_b^2 + l_c^2) > a^2 + b^2 + c^2$$

which is equivalent to prove

$$2\sum_{cyc}\left(bc - \frac{a^2bc}{(b+c)^2}\right) > a^2 + b^2 + c^2$$

On the other hand, we have

$$2\left(ab + bc + ca - \frac{4abc}{a+b+c}\right) - (a^2 + b^2 + c^2) = \frac{(a+b-c)(b+c-a)(c+a-b)}{a+b+c} > 0$$

It is sufficient to prove

$$2\sum_{cyc}\left(bc - \frac{a^2bc}{(b+c)^2}\right) > 2\left(ab + bc + ca - \frac{4abc}{a+b+c}\right)$$

$$\iff \frac{4abc}{a+b+c} > \sum_{cyc}\frac{a^2bc}{(b+c)^2}$$

$$\iff \frac{4}{a+b+c} > \sum_{cyc}\frac{a}{(b+c)^2}$$

which can be proved using the **Triangle inequality**

$$\sum_{cyc}\frac{a}{(b+c)^2} < \sum_{cyc}\frac{4a}{(a+b+c)^2} = \frac{4}{a+b+c}$$

as desired.

■

Solution 138.

If we note by a, b, c the sides of the triangle, then we have the following identity

$$(a^2 + b^2 - c^2)(b^2 + c^2 - a^2)(c^2 + a^2 - b^2) = 32s^2r^2(s - 2R - r)(s + 2R + r)$$

Therefore

$$s \geq 2R + r$$

This inequality is known as **Ciamberlini's inequality**.

Solution 139.

○ *Right inequality:*

We need to prove

$$a^2 + b^2 + c^2 \geq 8R^2$$

which is equivalent to

$$2s^2 - 8Rr - 2r^2 \geq 8R^2$$

$$\iff 2(s + 2R + r)(s - 2R - r) \geq 0$$

which is true by **Ciamberlini's inequality**.

○ *Left inequality:*

We will prove this inequality using two methods.

- *First method:*

By squaring $\overrightarrow{BC} = \overrightarrow{OC} - \overrightarrow{OB}$, we get

$$a^2 = 2R^2 - 2\overrightarrow{OB}.\overrightarrow{OC}$$

Or

$$2\overrightarrow{OB}.\overrightarrow{OC} = 2R^2 - a^2$$

Similarly

$$2\overrightarrow{OA}.\overrightarrow{OC} = 2R^2 - b^2$$
$$2\overrightarrow{OA}.\overrightarrow{OB} = 2R^2 - c^2$$

It follows that

$$9OG^2 = \|\overrightarrow{OA} + \overrightarrow{OB} + \overrightarrow{OC}\|^2$$
$$= 3R^2 + 2(\overrightarrow{OA}.\overrightarrow{OB} + \overrightarrow{OB}.\overrightarrow{OC} + \overrightarrow{OC}.\overrightarrow{OA})$$
$$= 9R^2 - (a^2 + b^2 + c^2)$$

Therefore

$$OG^2 = R^2 - \frac{1}{9}(a^2 + b^2 + c^2) \geq 0$$

as required.

- *Second method:*

We will prove the stronger inequality

$$a^2 + b^2 + c^2 \leq 8R^2 + 4r^2$$

Let

$$\begin{cases} x = b^2 + c^2 - a^2 \\ y = c^2 + a^2 - b^2 \\ z = a^2 + b^2 - c^2 \end{cases}$$

Thus, we need to prove that

$$x + y + z \le \frac{8a^2b^2c^2}{16S^2} + \frac{16S^2}{(a+b+c)^2}$$

Or

$$x + y + z \le \frac{(x+y)(y+z)(z+x)}{xy + yz + zx} + \frac{xy + yz + zx}{\left(\sum_{cyc} \sqrt{\frac{x+y}{2}}\right)^2}$$

Or

$$\left(\sum_{cyc} \sqrt{\frac{x+y}{2}}\right)^2 \le \frac{(xy + yz + zx)^2}{xyz}$$

which is true because

$$\left(\sum_{cyc} \sqrt{\frac{x+y}{2}}\right)^2 \le 3 \left(\sum_{cyc} \frac{x+y}{2}\right)$$

$$= 3(x + y + z)$$

$$\le \frac{(xy + yz + zx)^2}{xyz}$$

The inequality is proved.

■

Solution 140.

Let R, r, s, S be the circumradius, the inradius, the semiperimeter and the area of the triangle respectively.

• **First inequality:**

*Applying **Heron's formula**, we get*

$$S = \sqrt{s(s-a)(s-b)(s-c)}$$

But

$$S = sr = \frac{abc}{4R}$$

Therefore

$$r^2 = \frac{(s-a)(s-b)(s-c)}{s}$$

$$= \frac{s^3 - s^2(a+b+c) + s(ab+bc+ca) - abc}{s}$$

$$= -s^2 + (ab + bc + ca) - 4Rr$$

Consequently

$$ab + bc + ca = 4Rr + s^2 + r^2$$

Finally

$$ab + bc + ca - (4R^2 + 8Rr + 2r^2) = (s + 2R + r)(s - 2R - r) \geq 0$$

*which is true by **Ciamberlini's inequality.***

• *Second inequality:*

Similarly, we have

$$S = \frac{abc}{4R} = sr$$

Therefore

$$abc = 4sRr$$

and

$$abc - 4Rr(2R + r) = 4Rr(s - 2R - r) \geq 0.$$

*which is true by **Ciamberlini's inequality.***

■

Solution 141.

We will give two solutions to this problem.

• *First Solution:*

First of all, we have

$$
\begin{aligned}
\sum_{cyc} \frac{a^2}{b^2 + c^2} &= \sum_{cyc} \left(\frac{a^2}{b^2 + c^2} - \frac{1}{2} \right) + \frac{3}{2} \\
&= \sum_{cyc} \frac{2a^2 - b^2 - c^2}{2(b^2 + c^2)} + \frac{3}{2} \\
&= \sum_{cyc} \frac{a^2 - b^2}{2(b^2 + c^2)} + \sum_{cyc} \frac{a^2 - c^2}{2(b^2 + c^2)} + \frac{3}{2} \\
&= \sum_{cyc} (a^2 - b^2) \left(\frac{1}{2(b^2 + c^2)} - \frac{1}{2(a^2 + c^2)} \right) + \frac{3}{2} \\
&= \sum_{cyc} \frac{(a - b)^2 (a + b)^2}{2(a^2 + c^2)(b^2 + c^2)} + \frac{3}{2}
\end{aligned}
$$

*According to **Cauchy-Schwarz inequality**, we get*

$$(a^2 + c^2)(c^2 + b^2) \geq c^2 (a + b)^2$$

*By **AM-GM inequality**, we obtain*

$$c^2 = \frac{(c + a - b + c + b - a)^2}{4}$$

$$\geq (c+a-b)(c+b-a)$$

Therefore

$$(a^2 + c^2)(c^2 + b^2) \geq (c+a-b)(c+b-a)(a+b)^2$$

Consequently

$$\sum_{cyc} \frac{a^2}{b^2 + c^2} = \sum_{cyc} \frac{(a-b)^2(a+b)^2}{2(a^2+c^2)(b^2+c^2)} + \frac{3}{2}$$

$$\leq \sum_{cyc} \frac{(a-b)^2}{2(c+a-b)(c+a-b)} + \frac{3}{2}$$

$$= \frac{\sum_{cyc}(a+b-c)(a-b)^2}{2(a+b-c)(b+c-a)(c+a-b)} + \frac{3}{2}$$

$$= \frac{abc}{(a+b-c)(b+c-a)(c+a-b)} + \frac{1}{2}$$

$$= \frac{R}{2r} + \frac{1}{2}$$

$$\leq \frac{R}{2r} + \frac{R}{4r}$$

$$= \frac{3R}{4}$$

The problem is completely solved. Equality holds for $a = b = c$.

- *Second Solution:*

*Using **Ravi's substitution**: $a = x+y$, $b = y+z$ and $c = z+x$, we can rewrite the inequality in the following form*

$$\sum_{cyc} \frac{(x+y)^2}{(z+x)^2 + (y+z)^2} \leq \frac{3(x+y)(y+z)(z+x)}{16xyz}$$

or equivalently

$$\sum_{cyc} \frac{(x+y)^2}{(z^2+x^2)+(y^2+z^2)+2z(x+y)} \leq \frac{3(x+y)(y+z)(z+x)}{16xyz}$$

*According to **Titu's lemma**, we get*

$$\frac{(x+y)^2}{(z^2+x^2)+(y^2+z^2)+2z(x+y)} \leq \frac{1}{4}\left(\frac{x^2}{z^2+x^2} + \frac{y^2}{y^2+z^2} + \frac{(x+y)^2}{2z(x+y)} \right)$$

Therefore

$$\sum_{cyc} \frac{(x+y)^2}{(z^2+x^2)+(y^2+z^2)+2z(x+y)} \leq \frac{3}{4} + \sum_{cyc} \frac{x+y}{8z}$$

$$= \frac{\sum_{cyc} x^2(y+z) + 6xyz}{8xyz}$$

As such, it is enough to prove

$$\frac{\sum_{cyc} x^2(y+z) + 6xyz}{8xyz} \leq \frac{3(x+y)(y+z)(z+x)}{16xyz}$$

or equivalently

$$6xyz \leq \sum_{cyc} x^2(y+z)$$

which is obviously true. Equality holds for $a = b = c$.

∎

Solution 142.

*According to **Apollonius' theorem** and **AM-GM inequality**, we deduce*

$$4m_a\sqrt{a^2+b^2+c^2} = 2\sqrt{2b^2+2c^2-a^2}\sqrt{a^2+b^2+c^2}$$
$$\leq 2b^2 + 2c^2 - a^2 + a^2 + b^2 + c^2$$
$$= 3(a^2+b^2)$$

Therefore

$$m_a \leq \frac{3(b^2+c^2)}{4\sqrt{a^2+b^2+c^2}}$$

or equivalently

$$am_a \leq \frac{3a(b^2+c^2)}{4\sqrt{a^2+b^2+c^2}}$$

Similarly, we get

$$bm_b \leq \frac{3b(c^2+a^2)}{4\sqrt{a^2+b^2+c^2}}$$

$$cm_c \leq \frac{3c(a^2+b^2)}{4\sqrt{a^2+b^2+c^2}}$$

Summing up all these inequalities, we get

$$am_a + bm_b + cm_c \leq \frac{3\left(a(b^2+c^2) + b(c^2+a^2) + c(a^2+b^2)\right)}{4\sqrt{a^2+b^2+c^2}}$$

Therefore, we only need to prove

$$a(b^2+c^2) + b(c^2+a^2) + c(a^2+b^2) \leq a^3 + b^3 + c^3 + 3abc$$

*which is obviously true by **Schur's inequality**. Equality holds for $a = b = c$.*

∎

Solution 143.

Let R, r, s be the circumradius, the inradius, and the semiperimeter of the triangle respectively. Let

$$L = \sum_{cyc} ab(a + b)$$

We have

$$L = \sum_{cyc} ab(2s - c)$$
$$= 2s(ab + bc + ca) - 3abc$$
$$= 2s(4Rr + s^2 + r^2) - 12sRr$$
$$= 2s(-2Rr + s^2 + r^2)$$

Therefore

$$L - (16R^3 + 16R^2r + 12Rr^2 + 4r^3) = 2(s - 2R - r)(4R^2 + 2Rs + 2Rr + s^2 + sr + 2r^2)$$

According to **Ciamberlini's inequality**, we obtain

$$s \geq 2R + r$$

Finally

$$L \geq 16R^3 + 16R^2r + 12Rr^2 + 4r^3$$

as desired.

■

Solution 144.

Denote by $\triangle ABC$ our triangle, and let O and G be the cicrumcenter and centroid respectively. The triangle is acute so O and G are inside the triangle.

WLOG, we can assume that O is inside the triangle GAB. Let M and N be the projections of O and G on $[AB]$ respectively.

We can assume that $NA \geq NB$, hence

$$NA \geq \frac{AB}{2} = MA$$

We have

$$OA = \frac{MA}{\cos(\angle MAO)}$$

$$GA = \frac{NA}{\cos(\angle NAG)}$$

Thus

$$GA - OA \geq MA \left(\frac{1}{\cos(\angle NAG)} - \frac{1}{\cos(\angle MAG)} \right)$$

$\triangle ABC$ is an acute-angled triangle, therefore

$$\angle NAG \leq \angle MAG < \frac{\pi}{2}$$

Consequently

$$GA - OA \geq MA \left(\frac{1}{\cos(\angle NAG)} - \frac{1}{\cos(\angle MAG)} \right) \geq 0$$

On the other hand, we know that

$$\begin{cases} OA = R \\ GA = \frac{2m_a}{3} \end{cases}$$

Therefore

$$R \leq \frac{2m_a}{3}$$

The proof is complete. Equality holds for equilateral triangles.

∎

Solution 145.

Let S be the area of the triangle. We have

$$\begin{cases} h_a = \frac{2S}{a} \\ h_b = \frac{2S}{b} \\ h_c = \frac{2S}{c} \end{cases}$$

Thus

$$h_a h_b + h_b h_c + h_c h_a = 4S^2 \left(\frac{1}{ab} + \frac{1}{bc} + \frac{1}{ca} \right)$$

$$= r^2 \frac{(a+b+c)^3}{abc}$$

It is enough to prove

$$\frac{r(a+b+c)^3}{abc} \geq 8(R+r)$$

On the other hand, we know that

$$abc = 4RS = 2(a+b+c)Rr$$

Consequently, it is sufficient to prove

$$(a+b+c)^2 \geq 16R(R+r)$$

which is immediate by **Ciamberlini's inequality**. The proof is complete.

∎

Solution 146.

*Applying the **Triangle inequality** , we deduce*

$$\frac{a+b+c}{2} < a+b, b+c, c+a$$

Therefore

$$\frac{PQ}{a+b} + \frac{QR}{b+c} + \frac{RP}{a+c} < \frac{2}{a+b+c}(PQ+QR+RP)$$

$$< \frac{2}{a+b+c}[(CP+CQ)+(AQ+AR)+(BR+BP)]$$

$$= \frac{2}{a+b+c}[(CP+BP)+(AQ+CQAR)+(BR+AR)]$$

$$= \frac{2}{a+b+c}(a+b+c)$$

$$= 2$$

The proof is complete.

∎

Solution 147.

We will give three solutions to this problem.

● *First Solution:*

*According to **Cauchy-Schwarz inequality**, we obtain*

$$\frac{1}{a^2+b^2} + \frac{1}{b^2+c^2} + \frac{1}{c^2+a^2} \geq \frac{9}{2(a^2+b^2+c^2)}$$

Therefore, we only need to prove

$$(a+b+c)^3 \geq 18\sqrt{3}(a^2+b^2+c^2)r$$

*The function $x \to \tan\left(\frac{x}{2}\right)$ is convex for $x \in [0, \pi]$, thus according to **Jensen's inequality***

$$\tan\left(\frac{A}{2}\right) + \tan\left(\frac{B}{2}\right) + \tan\left(\frac{C}{2}\right) \geq 3\tan\left(\frac{A+B+C}{6}\right)$$

$$= 3\tan\left(\frac{\pi}{6}\right)$$

$$= \sqrt{3}$$

Consequently, it is sufficient to prove

$$(a+b+c)^3 \geq 18\left(\sum_{cyc}\tan\frac{A}{2}\right)(a^2+b^2+c^2)r$$

Furthermore, we know that

$$\tan\left(\frac{A}{2}\right) + \tan\left(\frac{B}{2}\right) + \tan\left(\frac{C}{2}\right) = \frac{4R+r}{s}$$

Thus, it is enough to prove

$$(a+b+c)^3 \geq 18\left(\frac{4R+r}{s}\right)(a^2+b^2+c^2)\,r$$

*By **Euler's inequality**, we get $R \geq 2r$. Thus, we only need to prove*

$$(a+b+c)^3 \geq 18\left(\frac{4R+\frac{R}{2}}{s}\right)(a^2+b^2+c^2)\,r$$

or equivalently

$$(a+b+c)^5 \geq 81\,abc\,(a^2+b^2+c^2)$$

which is straightforward as follows

$$\begin{aligned}
(a+b+c)^5 &= \frac{(a+b+c)^6}{(a+b+c)} \\
&= \frac{(a^2+b^2+c^2+2ab+2bc+2ca)^3}{(a+b+c)} \\
&\geq \frac{27(ab+bc+ca)^2(a^2+b^2+c^2)}{a+b+c} \\
&\geq \frac{81abc(a+b+c)(a^2+b^2+c^2)}{a+b+c} \\
&= 81abc(a^2+b^2+c^2)
\end{aligned}$$

as required. Equality holds for $a = b = c$.

- ***Second Solution:***

*After using **Ravi's substitution**, we need to prove that*

$$\sum_{cyc}\frac{1}{x^2+y^2+2z^2+2xz+2yz} \geq \frac{81\sqrt{3xyz}}{8\sqrt{(x+y+z)^7}}$$

*Using **Cauchy-Schwarz inequality**, we deduce that*

$$\left(\sum_{cyc}\frac{1}{x^2+y^2+2z^2+2xz+2yz}\right)\left(\sum_{cyc}(x^2+y^2+2z^2+2xz+2yz)\right) \geq 9$$

Thus, it is enough to prove that

$$4(x+y+z)^7 \geq 243xyz\left(\sum_{cyc}(x^2+xy)\right)^2$$

WLOG, we can assume that $x + y + z = 1$. *Let* $t \in [0, 1]$ *such that* $xy + yz + zx = \frac{1-t^2}{3}$.
According to **Theorem 39**, *we get that*

$$xyz \leq \frac{(1-t)^2(1+2t)}{27}$$

Therefore, we only need to prove

$$f(t) = 4 - (1-t)^2(1+2t)(2+t^2)^2 \geq 0$$

On the other hand, we have

$$f'(t) = 2t(1-t)(t^2+2)(7t^2 - 2t + 4) \geq 0$$

Consequently, f is increasing on $[0, 1]$, and we get

$$f(t) \geq f(0) = 0$$

The proof is complete. Equality holds for $a = b = c = 1$.

- ***Third Solution:***

After using **Ravi's substitution**, *we need to prove that*

$$\sum_{cyc} \frac{1}{x^2 + y^2 + 2z^2 + 2xz + 2yz} \geq \frac{81\sqrt{3xyz}}{8\sqrt{(x+y+z)^7}}$$

Using **Cauchy-Schwarz inequality**, *we get*

$$\left(\sum_{cyc} \frac{1}{x^2 + y^2 + 2z^2 + 2xz + 2yz} \right) \left(\sum_{cyc} (x^2 + y^2 + 2z^2 + 2xz + 2yz) \right) \geq 9$$

Thus, it is enough to prove that

$$4(x+y+z)^7 \geq 243xyz \left(\sum_{cyc} (x^2 + xy) \right)^2$$

which is equivalent to $f(w^3) \geq 0$ where f is a decreasing function. According to **uvw** *method, it is enough to prove the last inequality for the maximal value of $w^3 = xyz$, which happens for equality case of two variables.*

WLOG, Let's assume $y = z = 1$. In this case, we obtain

$$(x-1)^2(512x^5 + 629x^4 + 518x^3 + 217x^2 + 64x + 4) \geq 0$$

which is obviously true. Equality holds for $a = b = c = 1$.

■

Solution 148.

According to **Tereshin-Belabess inequality,** *we obtain*

$$m_a \leq \frac{(b+c)^2}{16r}$$

or equivalently

$$\frac{m_a}{b+c} \leq \frac{b+c}{16r}$$

Similarly, we get

$$\frac{m_b}{c+a} \leq \frac{c+a}{16r}$$

$$\frac{m_c}{a+b} \leq \frac{a+b}{16r}$$

Summing up all these inequalities, we get

$$\frac{m_a}{b+c} + \frac{m_b}{c+a} + \frac{m_c}{a+b} \leq \frac{a+b+c}{8r}$$

The proof is complete. Equality holds for $a = b = c$.

Comment 32.

We will show here how to prove **Tereshin-Belabess inequality.**

○ **Proof:**

Using **Apollonius' theorem,** *we get*

$$m_a = \frac{1}{2}\sqrt{2(b^2+c^2) - a^2}$$

Let s *be the semi-perimeter. According to* **Heron's formula,** *we have*

$$r = \sqrt{\frac{(s-a)(s-b)(s-c)}{s}}$$

Therefore, we need to prove

$$(b+c)^4(a+b+c) \geq 16(2(b^2+c^2) - a^2)(a+b-c)(b+c-a)(c+a-b)$$

Using **Ravi's substitution:** $a = x+y$, $b = y+z$ *and* $c = z+x$, *we can rewrite the inequality as follows*

$$(x+y+z)(x+y+2z)^4 \geq 64xyz\left(2((z+x)^2 + (z+y)^2) - (x+y)^2\right)$$

WLOG, we can assume $z = 1$. *Let* $x+y = 2u$ *and* $xy = v^2$. *We have* $u \geq v$ *and the inequality is equivalent to*

$$f(u) = (1+2u)(2+2u)^4 - 64v^2(4+4u^2 - 4v^2 + 8u) \geq 0$$

On the other hand, we have

$$f'(u) = 32(1+u)(3+11u+13u^2+5u^3-16v^2)$$
$$\geq 32(1+u)(3+11u-3u^2+5u^3)$$
$$\geq 0$$

f is an increasing function, therefore

$$f(u) \geq f(v)$$
$$= 16(v-1)^2(2v^3+13v^2+8v+1)$$
$$\geq 0$$

Finally, we have proved our lemma. Equality holds for $a = b = c$.

■

Solution 149.

*According to **Apollonius' theorem** and **AM-GM inequality**, we get that*

$$\frac{m_a m_b}{b^2+c^2} = \frac{\sqrt{2b^2+2c^2-a^2}\sqrt{a^2+b^2+c^2}\sqrt{2a^2+2c^2-b^2}\sqrt{a^2+b^2+c^2}}{4(a^2+b^2+c^2)(b^2+c^2)}$$
$$\leq \frac{(2b^2+2c^2-a^2+a^2+b^2+c^2)(2a^2+2c^2-b^2+a^2+b^2+c^2)}{16(a^2+b^2+c^2)(b^2+c^2)}$$
$$= \frac{9(b^2+c^2)(a^2+c^2)}{16(a^2+b^2+c^2)(b^2+c^2)}$$
$$= \frac{9(a^2+c^2)}{16(a^2+b^2+c^2)}$$

Similarly

$$\frac{m_b m_c}{c^2+a^2} \leq \frac{9(b^2+a^2)}{16(a^2+b^2+c^2)}$$

$$\frac{m_c m_a}{a^2+b^2} \leq \frac{9(c^2+b^2)}{16(a^2+b^2+c^2)}$$

Summing up all these inequalities, we get

$$\sum_{cyc} \frac{m_a m_b}{b^2+c^2} \leq \sum_{cyc} \frac{9(a^2+c^2)}{16(a^2+b^2+c^2)} = \frac{9}{8}$$

The proof is complete. Equality holds for $a = b = c$.

■

Solution 150.

We will use the following well-known lemma: In any triangle with side lengths a, b, c and inradius r, the following inequality holds

$$r \leq \frac{\sqrt{3}(a+b)^2(a+b+c)}{36(a^2+b^2)}$$

Using the lemma, we get

$$\frac{a^2+b^2}{a+b} \leq \frac{\sqrt{3}(a+b)(a+b+c)}{36r}$$

Similarly

$$\frac{b^2+c^2}{b+c} \leq \frac{\sqrt{3}(b+c)(a+b+c)}{36r}$$

$$\frac{c^2+a^2}{c+a} \leq \frac{\sqrt{3}(c+a)(a+b+c)}{36r}$$

Summing up all these inequalities, we obtain

$$\frac{a^2+b^2}{a+b} + \frac{b^2+c^2}{b+c} + \frac{c^2+a^2}{c+a} \leq \frac{(a+b+c)^2}{6\sqrt{3}r}$$

This ends the proof. Equality holds for $a = b = c$.

Comment 33.

We will show here how to prove:

$$r \leq \frac{\sqrt{3}(a+b)^2(a+b+c)}{36(a^2+b^2)}$$

○ ***Proof:***

*Using **Ravi's substitution:** $a = x+y$, $b = y+z$ and $c = z+x$, we can rewrite the inequality as follows*

$$\frac{(x+2y+z)^4(x+y+z)^3}{108xyz(x^2+2y^2+z^2+2xy+2yz)^2} \geq 1$$

Let $x^2 + 2y^2 + z^2 + 2xy + 2yz = 2m$ and $xy + yz + 2zx = n$. We have

$$(x+y+z)(x+2y+z) = n+2m$$

*According to **AM-GM inequality**, we have*

$$(n+2m)^3 \geq 27nm^2$$

Therefore

$$\frac{(x+2y+z)^4(x+y+z)^3}{108xyz(x^2+2y^2+z^2+2xy+2yz)^2} = \frac{(x+2y+z)(n+2m)^3}{432xyzm^2}$$

$$\geq \frac{(x + 2y + z)n}{16xyz}$$

$$= \frac{(x + 2y + z)(xy + yz + 2zx)}{16xyz}$$

$$\geq \frac{4\sqrt[4]{xy^2 z} \cdot 4\sqrt[4]{x^3 y^2 z^3}}{16xyz}$$

$$= 1$$

The proof is complete. Equality holds for $a = b = c$.

www.ingramcontent.com/pod-product-compliance
Lightning Source LLC
Chambersburg PA
CBHW081719220526
45468CB00008B/1901